# Changes and Challenges
## A life in coal (and beyond)

### Trevor Massey

*Challenges are what make life interesting;
overcoming them is what makes life meaningful.*

Editing, design and typesetting by Mosaic Design and Print,
Mosaic (Teesdale) Ltd. 01833 640893 www.mosaicteesdale.co.uk.

# Foreword

An only son who spent his early years surrounded by older relatives. Not the background to produce a highly motivated young man who would be a go-getter in the world! Rather, a shy, reserved lad of limited self-confidence. He did, though, benefit from good teaching at his junior school which got him a place a the fine Queen Elizabeth Grammar School in Wakefield. There, the competition was challenging and he was always in the second division of the student hierarchy. His university ambitions were thwarted by his inability to get basic credit grades in General Science and Latin. This led to tears and despair at the prospects for the future. Surely, this would suggest a hopeless subject for a biography?

However life is full of surprises and sometimes the confluence of people and events lead to dramatic, life-changing decisions. Who would have thought that an underground visit to Woolley Colliery's Parkgate seam on a Saturday morning would have led to a decision to enter the mining industry? It invalidated all his past education in classics and arts subjects and demanded he tackle the science and maths subjects that he feared. His mentor, H J Atkinson, Area General Manager in the Number 6 Area of the Yorkshire Coalfield, gave clear objectives: he must go to university and get a degree in Mining Engineering. In retrospect that decision to enter the coal industry by a young man was a monumental risk that could have led to disaster.

The first step was fundamental and irreversible: leave the Grammar School and start as a workman at a mine and study science subjects at a night school. The fall from status was complete, as he had to travel to the mine on a bicycle and return home in his pit clothes as there were no spare lockers in the pit-head baths. The next three years demonstrated that any subject can be mastered if the motivation is right.

Getting a good degree and the experience to obtain a Mine Managers Certificate was not a passport to success. There were many men with those qualifications in the industry who spent their whole career in junior management roles. But his career was peppered by changes and challenges. Every time he made a success of one post he was offered a bigger challenge. Every change meant that he had to expand his expertise to solve the varied problems of each appointment.

Eventually the posts were outside the conventional mine management roles. The Selby Project introduced challenges of stretching the mining technology to new levels of performance way above those in use anywhere in the world. It also required financial management of operations such that very large sums of money (£3.7 million per week) achieved the planned performance of work completed.

In the same way, managing the Technical Department of British Coal required the motivation of brilliant brains to forge new equipment and operating systems to make the mining of coal safer and more efficient. At the same time the size of the department had to be reduced to match the reducing size of the mining operations. There were similar requirements in the Purchasing and Stores Department.

Was this life and experience of interest to warrant writing it up into an autobiography? Some people said that it might be. The first request was a demand by one of my sons that we needed to record the history of our family otherwise it would be lost for ever. When that was completed it was accepted that a first objective had been achieved. The remainder of the story was written when he was persuaded to give it priority over other writing he was working on. A comment by his wife was succinct and pointed: "Write about what you know instead of other occupations and people".

As I tackled this writing task I was beginning to realise that the coal industry might be in terminal decline. In my last year in the industry, 1992, I was President of the Institution of Mining Engineers. Some of the speeches I made in that year identified the symptoms of pending death due to political decisions being applied to the industry. Parts of these speeches are included in the appendix, which identify details of these political decisions.

To be fair to my life story, a lot has happened since my retirement 26 years ago. To complete the story, it seemed appropriate to include examples of activities that I have been involved in during that time.

When I entered the coal industry in 1951 it was inconceivable to dream that there might be the end of coal mining in the UK during my lifetime. But it has happened, so my experiences reflect records of aspects in the technical and social history of a great industry.

# Contents

| | | |
|---|---|---|
| Chapter 1 | Family history; kids! | 1 |
| Chapter 2 | A new Massey family | 9 |
| Chapter 3 | Growing up; an early lesson in coal mining; a family sing-song | 14 |
| Chapter 4 | A new world; Queen Elizabeth Grammar School, Wakefield 1945-51 | 35 |
| Chapter 5 | Major changes at home; farewell, Grandad | 38 |
| Chapter 6 | Further education and two big decisions | 44 |
| Chapter 7 | A miner starting at the bottom, 1951-54; working underground | 49 |
| Chapter 8 | A new henhouse and a new church; route to a wife | 55 |
| Chapter 9 | University route; a university place — just?; Directed Practical Trainee, 1958/59 | 62 |
| Chapter 10 | First step on the ladder; shaft sinking; Colliery Overman, Wharncliffe Woodmoor 123, 1960-61 | 70 |
| Chapter 11 | First steps in senior management, 1961-65; back to where it all started, Woolley Colliery, 1963 | 80 |
| Chapter 12 | A new mine and a big mine, 1969-72; official opening ceremony of Riddings Drift Mine; General Manager, South Kirkby Colliery, 1971-72 | 95 |
| Chapter 13 | Production Manager, South Yorkshire Area, 1972-74 | 112 |

| | | |
|---|---|---|
| Chapter 14 | Doncaster Area, Chief Mining Engineer, 1974-77 | 118 |
| Chapter 15 | South Yorkshire Area, Deputy Director Mining, 1977-81; A home of our own; an added responsibility; George Hayes' accident | 125 |
| Chapter 16 | Deputy Director Mining, Selby, 1981-85; flooding, 1982; President of Midland Institute of Mining Engineering; World Mining Congress, India, November 1984 | 144 |
| Chapter 17 | A move to Headquarters, 1985-91; Technical Department organisation; examples of R&D challenges; legislation; other people's problems; an evaluation by the Boss; representation on external committees; the right organisation for the coal industry; an additional role; a board meeting at Bretby; another idea to exploit Bretby's facilities | 175 |
| Chapter 18 | Head of Supply and Contracts Department, April 1991-November 1992; overview; opportunities for change; the challenge of 1992; relocating the department from Coal House; the Annual Conference of the Institution, Blackpool 1992; a major shock for the coal industry; aspects of retirement; the investiture; a quiet goodbye. | 217 |
| Chapter 19 | Scenes from retirement | 250 |
| Appendix | Speech in Cardiff, 19 March 1992; speech in Sheffield, 20 March 1992; speech in Scotland, 21 March 1992 | 275 |
| Glossary of technical terms | | 287 |

# Chapter 1

*Family history*

History, as taught at school, is a catalogue of national and international events over the years. Family history is often anecdotal snippets about the lives of former family members. When the two streams of information are put together the progress of a family through those historical times is often remarkable.

My grandparents were born in the 1870s. It was the age of the industrial revolution in the UK; the age of steam power and the railway expansion. It was the time of expanding towns and cities to house the workers for the factories that were being built. The move of workers from the country to the towns was continuous. My grandparents were involved in the industrial revolution and they moved locations to better their prospects.

The industrial revolution was built on coal. Coal provided the power and to get the coal, new technologies had to be developed to gain access to the coal seams. Shafts had to be sunk as the deeper seams were worked. Steam engines were developed to pump water out of the mines. Steam engines were then adapted to wind the coal up the shafts. Steam engines were then modified to transport the coal from the pits to the markets; thus, began the start of the railway revolution. Great Britain led the world in the production of coal. More coal was produced in the UK than in any other country in the world until the 1880s when the USA became the world leader. At the end of the 19th Century one in ten of the total labour force in the UK was in coal mining. It is not surprising, therefore, that my forefathers were miners.

During the early years of my grandparents' lives there was instability in Europe. The Franco-Prussian war in Europe had a lasting effect on relations between Germany and France. As far as the UK was concerned the battles were mainly in defence of the Empire or expanding it into new territories. The British navy ruled the world and the UK was dominant in international trade.

In the new 20th century the UK was dragged into the wider conflicts in the world. The experience in the Boer war was a painful victory, not a confident curtain-raiser for the First World War. The whole country was traumatised by the scale of the operations of the First World War. The recruitment of men and horses was recalled by my parents. They described the horses being rounded up and then led away by the army. The fact that so many men, and virtually all the horses, never returned affected a great number of families in the UK.

The post-war period introduced economic problems. It is noted for the industrial unrest in the coal mines which led to the 1926 strike. That strike embittered relationships between the miners and the mine owners that were never resolved. The world recession in output and trade in the 1930s removed any chance of job security for most workmen. Unemployment was not underpinned by state social security. It was not the time to grow a family; there was always the threat of destitution. That was the scenario in which the Massey and Hemingway families grew up.

Let me deal with the Hemmingway family first. My grandfather Newman, who died before I was born, had the misfortune to lose his father when he was eight months old and he was brought up by his aunt in Leeds. His father was in a senior position in a woollen mill, but when he died the family were left destitute. When he grew up, my grandfather moved back to Batley to be with his mother, and he started working in a woollen mill. In the 1890s, work was very bad in the woollen mills, so Newman and his brother, Cyrus, set out to walk until they found work. They ended up in the village of Higham, near Barnsley and were offered work in the local mine, so they decided to settle there. Later, they lived in a cottage near the

Methodist chapel and they fetched their mother (Sarah Ann) over to live with them, and she became caretaker of the chapel.

Newman Hemmingway met his wife, Grace Birtles, and married her when he was 28 years old. I remember Grace. She was a feisty character who dominated her children with her passion for cleaning and polishing the house. Her theory was that if you kept your children working in the house or garden they would not get into trouble. How much Grace's attitudes affected her husband I don't know, but he developed a busy life outside his family. He became a charge man on a coal face at Woolley Colliery. This was the time when the charge man drew all the money for the coal face each week and shared it out among the team — some responsibility! At home, he had two allotments and kept hens and grew vegetables for the family. He also had a herb garden and made cures for ailments which he gave to people who could not afford to go to the doctor. But then there was the church. He was a faithful Christian. In the Centenary Souvenir for Higham Methodist church he is recorded as Trustee, Sunday School Superintendent and Church Secretary for thirty years; he was also President of the Society. He died of stomach cancer in 1932, aged 56 years. I have no doubt that he played a part in the scheme to have a new Methodist church built in Higham, but he didn't live to see the stone-laying on 4th May 1935, and the official opening on 29th February 1936. As far as my mother was concerned, he was her hero and he passed on to her some of his characteristics which mellowed those she received from her mother. My mother was the eldest of six children, five of whom survived to become adults. So, the Hemmingway family was established when Newman died; it was on a far better basis than might have been expected when he set out from Batley to seek work.

The Massey family also grew and prospered to a degree that was unpredictable at the start of the 20th century, but there are two mysteries in this history that are difficult to understand.

My Grandfather John Thomas Massey was born in January 1872, in Biddulph in Staffordshire. His father and his grandfather were coal miners. His father, also called Thomas, died when he was 49, in 1884. The next record we have is a report in the local newspaper about Grandad on the occasion of his Golden Wedding in 1942. It reports that he lived in Monk Bretton as a boy, and worked at the Craig's pit, in Smithies, as a pony driver. At that time, he lived with his mother and he was the sole support for her. The report refers to the fact that, in the winter in those days, the boys in the mines never saw daylight from Monday to Friday. In the 1891 census Grandad and his mother had moved to Higham and she was taking in boarders as well as her family. This must have been short-lived, because she died in February 1892, back in a Red Cross hospital in Biddulph, of gastric ulcer and exhaustion.

Grandad Massey's life moved to a new phase after the death of his mother. In November 1892, he got married to Mary Hallworth. She was born in Hazel Grove, in Cheshire, on 27th March 1875. A major question arises: how did Grandad, from Staffordshire, manage to meet Mary, from Cheshire? There is a clue in the 1891 census. The Hallworths had moved to Higham, but there is no reference to their daughter Mary in the census record. Could Grandad have met her when she visited her parents in Higham? It is a possibility. They got married in Salford, in Manchester. They both gave the wrong ages on the wedding certificate, claiming to be 21 years old. In fact, Grandad was 20 and Grandma was only 17 years old. There might have been a sense of urgency with the wedding, because their first child, Thomas, was born in 1893.

Grandma Massey, we know, was, for a period, a hat maker. She said that her job was to put the leather lining around the inside of bowler hats. She had trouble with her hands sweating, which marked the leather, so she was dismissed. She obviously stayed behind in Cheshire when her parents moved to Yorkshire.

When they were married, Grandad and Grandma settled down in Higham in an Old Silkstone Colliery Company house. The initial house was in the New Row but they later moved to the Concrete Row. In all, they spent 40 years in colliery company houses. Grandad worked in the Silkstone seam of the Old Silkstone Colliery Company at Church Lane Dodworth. It would be a 15-minute walk over the fields to the pit each day.

I am under the impression that it was while they lived on the Concrete Row that Grandad first started gardening in a significant way and keeping livestock. It is possible that he ceased being a miner after the 1926 strike, and concentrated on his smallholding activities and gardening work. He did gardening work for a house in Cawthorne and then for a house near Higham pit. Around 1930, he made a move to a new location in Higham, by buying over three acres of land. The piece of land was at the southern edge of Higham, where it merged into Barugh Green. How did he manage to finance this in the major world slump of that period? There is no answer to that question. By selling on building plots for four houses he would have been able to recover some of his investment. In partnership with my father they built two semi-detached houses on the site.

Back to Grandma and Grandad Massey.

### Kids!!

Grandma Massey had relatives in Hazel Grove, Cheshire, who were in the distribution trade to shops for cleaning materials. They had a place (it could not be classed as a factory), in the 1940s, where they bottled detergent and other household products for distribution to retailers. They also had connections with the pottery industry and they may have traded this side of their business through a shop, but I never remember seeing one. They must have been reasonably wealthy because they had a car when cars were a rare sight. They used to come over from Hazel Grove and bring Grandma's sister, Aunt Jinny, with them. Aunt Jinny lived in Portwood, in the middle of Stockport. I think their first visits might have started before the Second World War, when I was a very small boy.

The fellow who drove the car was called Frank Hallworth. My mother described him as a sloppy fellow, because he always insisted on kissing all the ladies when he arrived on a visit and again when he departed. Frank and his wife always brought gifts for Grandma, and they took garden produce and eggs with them on the return trip. Grandma once commented to my mother, when they were discussing one of Frank's visits after he had driven away, 'Frank is still the same as he's always been, kissing all the ladies,' she said.

'He never changes. That's something our Dad has never done with me. He has never ever kissed me.'

This was a remarkable statement. It represented a definition of standard practice in a marriage that had lasted nearly fifty years. My Grandad was a man who did not show his emotions freely, but he was not a fierce or hard man and he could be good company. However, if Grandma was to be believed, kissing was not in his repertoire. I can never remember him kissing any relatives, or friends, or his children. There is a wedding photograph of one of their sons; Grandma is on the photograph but there is no sign of Grandad.

But Grandad Massey must have been a passionate man because he fathered twelve children with Grandma, over a 20-year period. There was a hypnotic time sequence about the births. Every two years, from 1893 to 1913, Grandma delivered a new child. She also had one set of twins, in 1909. This would suggest a family of different ages growing up together, the young ones being supported by the older ones. A growing family with babies fed in at one end and growing up to be children, and then, as adults, getting married as they matured and moving away. But it was not like that at all. The family was steeped in tragedies with the children.

When Jane Elizabeth, born in 1897, was aged two, she reached up to the mantelpiece over the fire and her nightdress caught fire and she was burned to death.

Fred Massey was born as one of the twins in 1909, but died at the age of three months.

James Massey, the youngest child, born in 1913, fell off the roof of an outbuilding near his home, in 1918, and died aged five years.

1918 was a particularly bad year for the family, as there were two other deaths. There is a story told that the family attended the funeral of one of their children and when they got back home another son had died.

The eldest son, Thomas, died aged 26 in the flu epidemic in 1918. He was married and had two children. Arthur Massey, born in 1899, also died of the flu epidemic in 1918, when he was 18 years old.

All these deaths occurred when the family lived on Concrete Row in Higham. Grandma never discussed these tragedies, but she did describe how tough it was living with a large family at that time. She explained how she walked the two miles to Barnsley each Saturday, with four large shopping bags. She went to the market when it was near closing time and picked up as much food as possible at discounted prices. She described the walk back from Barnsley, with the full bags, as a major physical challenge. On many of those walks she would have been pregnant. She also talked of starving herself from time to time, until she felt faint, to allow the kids to eat.

My mother was shocked when she visited the house in Concrete Row, after she started courting Roland, my Dad, to find that only Grandad and Grandma sat at the table. All the children stood at the side of the table to eat their meals.

Money was always tight for the Massey family, and there was no escaping through education to a better way of life. The village school was at Barugh Green, which was a mile walk for the children of Concrete Row. At the Barugh Green school, Charles passed the scholarship to go to the Grammar School in Barnsley, but Grandad ruled that the family could not afford the additional cost, and he would have to go down the mine. He never forgave Grandad for that decision.

But there were additional benefits in the education process. The final school that they attended until they were aged 13 was Higham School. The headmaster, Mr Woodcock, was very keen on music and he taught the scholars choral singing. He taught them to sing in four-part harmony and he used the tonic-sol-fa method. They could sing the choruses from the Messiah in tonic-sol-fa. This had a major impact on the Methodist Churches, providing a source of accomplished singers. I remember the choir at Higham Methodist Church having over 30 members and putting on singing events as well as supporting the Sunday services. On one occasion they performed a version of the opera Carmen. At the end of a long rehearsal the conductor, headmaster Lawrence Naylor, suddenly remembered that they had to sing the chorus, 'And the Glory,' from the Messiah at the Sunday night service. The choir just gathered together in the schoolroom and sang that chorus, without any music. I will never forget the singing that night; it was as exciting as any performance I have ever heard.

*Grandma and Granddad with Judy on the lawn, in about 1938.*

# Chapter 2

*A new Massey family*

My mother and dad's wedding was dependent on the completion of the new house being built in Higham. The house had an attractive name — Sunnymead — but it was not a finished product when they moved in. For years, my Dad made modifications and additions when finance allowed him. Their wedding was delayed because Grandad Hemmingway was very ill, and my mother insisted on delaying the wedding until after he had died. She wanted to be with him as much as possible, and she felt the need to support her mother and her younger brothers and sisters through this family tragedy. He died on 27th September 1932, aged 56.

The wedding of my parents took place on 10th December 1932, at Blucher Street Methodist Church, in Barnsley, because Higham Chapel was not licensed for weddings. The wedding photographs show a couple who were stepping out into the unknown and they knew it. There was a look of fear, tempered with hope, on their faces. They had gone through the challenge of building their own house alongside Grandad Massey's house. The houses might be built, but there was no guarantee that the relationships between the two families would be easy as life progressed, and the families went their separate ways.

*Roland and Vera, my parents, on their wedding day*

A lot of the basic site work for the houses was done by the family, and the final cost was a little over £200 each. In the village, the development was regarded as high risk and there was a view that both families would go bankrupt. This led to the famous row of egg cups that my mother had on a shelf in one of her kitchen cupboards. In the egg cups, when my Dad brought home his weekly wage, she shared out the essential money that they had to pay to cover the house mortgage and the living costs. At times, it would have been a balancing act as my Dad was only on a small wage as a property repairer for the colliery company houses in Dodworth and Higham. He supplemented his earnings by other jobs: he collected the rents for his brother-in-law in Barnsley, and he also did repairs on the same houses; he also did repairs on other houses in Higham. He was the local agent for a Friendly Society, collecting subscriptions and paying out benefits. My mother also let out a room in the house for the doctor from Dodworth to hold his local surgeries. They didn't go bankrupt, but I recall a period in the Second World War when my Dad badly broke his leg at work. The household financial reserves were nearly dried up by the time his compensation came through.

My mother developed a careful approach to money that lasted all her life. She only spent on luxuries when she had saved up to buy them. She cooked plain food in the same, open fire, coal oven for over 65 years, always using vegetables and fruit from the garden. Her menus were not adventurous, but she maintained a healthy, active life until her early 90s.

As was the custom in those days, when a woman got married she had to give up her job. My mother gave up her job in a shirt factory in Barnsley. So, her life was centred around her house and the smallholding. But it was not long before there was an extension of the family.

I was born on 3rd May 1934, and christened Charles Trevor Massey. I was to remain an only child, except for a short period when we had an evacuee from London during the period of the flying bombs near the end of the Second World War.

My memories of early life are sparse. I recall going in my pram to a baby clinic in Gawber, a village about a mile from Higham. Obviously, I spent a lot of time with my Grandad in the smallholding, watching his life with the pigs and hens and his garden. The opportunity to mix with other children of my age was limited, which might have been a blessing, because the council house estate near our house was noted for its rough families. Some of the parents were good at drinking, and their houses lacked curtains and any sense of tidiness. Their gardens were scrapyards of broken prams and bicycles. Both parents and children rated themselves by their ability to fight their way out of any problems in life. When I started school in Barugh Green, I had to walk past this estate four times each day — morning, lunchtime twice and evening. I kept to the opposite side of the road, but I was accosted on occasions by bullies from the estate. In one respect I envied some of the children. Their evening meal was often a thick slice of bread covered in jam. They ate this wandering about on foot. I was going home to a cooked meal each evening, when my Dad came home from work, but the jam and bread often looked more desirable.

My Grandad had an Alsatian dog called Judy. She was normally a friendly dog, but if another dog, going up the road, decided to enter the drive, she became the ultimate guard dog. She would charge blindly along the drive to tackle the intruder. On one occasion I was in her line of travel and I was tossed in the air and fell on my head on the concrete path and was knocked unconscious. I still have a lump on my forehead. Such an accident now would involve brain scans and hospital investigations, but I was just put in a darkened bedroom for a few days and observed for any side effects. Luckily there have been none.

The other thing I remember was the Methodist church in Higham. My father was Sunday School Superintendent and my mother played the piano for the Sunday School. Both my father and mother sang in the choir so there were choir practices each week, and special concerts as well as the Sunday services. That church was an essential part of our lives for the first 19 years of my life.

The Barugh Green Junior Mixed and Infant School was a major influence on many children of that locality. The Headmistress was Miss Marshall, a dominating woman who had clear objectives for her staff, the pupils and their parents. My mother once went to see her about me when I was not too happy at school. Mother was put in her place by a simple statement, 'Surely you want your child to achieve more than his parents.' There is only one answer to that question.

At the end of the six years the class took the famous 11-plus exams. We also took an additional exam for the George Beaumont Scholarship. This was a foundation to support students so that they could go to the Queen Elizabeth Grammar School in Wakefield rather than the local grammar school in Barnsley. There were over 120 students took the examination in Darton, for about eight scholarships that year. In one of my rare moments of supreme confidence I announced, when I got home, that I thought I had been successful. I was right and that opened up six years of education that affected my life, but not my career. More of that later.

# Chapter 3

*Growing up; the domino championship*

Grandad Massey liked to play dominoes in the evenings, but the games could only start when he had completed his evening washing routine and he'd got changed after his day in the garden. He had a large market garden of nearly three acres, growing fruit and vegetables that he sold at the door and through a local greengrocer.

Grandad also kept pigs and hens throughout the Second World War. The rule brought in at the start of the war was that smallholders were allowed to keep the same numbers of stock that they had in 1938. Grandad's smallholding was well stocked in 1938, so he had a thriving business during the war. Looking after the hens and pigs, as well as the garden, involved some dirty work, and so his evening routine was thorough and extensive. It all took place at the sink in the corner of the kitchen.

The routine started with him lowering his trousers, which had to be changed as they were dirty and often carried a pungent smell from the pigs. It might be thought that lowering his trousers in the public arena of the kitchen could be offensive to the family members watching, but that wasn't the case. He was not a tall man — about five feet six inches — and he wore long, voluminous shirts. As he lowered his trousers, his shirt unfolded like a theatre curtain to well below his knees. He then started on his toilet. It began with a shave of the grey bristles on his face, followed by a good wash of arms, hands, face and often his hair. When he had given himself a good rub with a towel his appearance was vastly changed. Then a different pair of trousers was pulled up and the shirt tucked in. His final task was to brush his hair and moustache and put a wave in his hair. His next move was to sit in his chair in the corner of the kitchen. He was an impressive sight as the chair raised him up and he looked resplendent — all shining, clean and tidy. He often then took out his clay pipe and had a smoke.

Like many pipe smokers Grandad had the need to spit out to clear his mouth from time to time. In executing this he demonstrated a remarkable skill commensurate with the finest bowlers on the cricket field. His aim was for the spit to land in the open coal fire — a distance of around four feet. The problem was that there was usually a kettle on the bar simmering for the next cup of tea. Grandad had the ability to project the spit so that it bent around the kettle and landed with a sizzle in the open fire. When he had rested and enjoyed his smoke he instructed that the dominoes should be brought out onto the kitchen table.

The number of people playing dominoes varied, but in 1944 we had an additional member of the family as we had an evacuee from London at the time of the flying bombs. Her name was Avril and she took part in a memorable domino championship. Avril was partnered with Grandad and I was partnered with my Dad. The winner would be the first pair to win seven games. The winner of each game took a card from a pack of cards and kept it on display to show their score. It was a long and close contest and the results went from one team to the other until the score was six games each. So, everything depended on the outcome of the final game. That was also close, and all the dominoes were played except one held by me and one held by Avril. It was my lead and I was sure that I had the winner; I put my domino down, which was the blank/one and triumphantly shouted: 'Beat that!'

Avril put her domino down —the double blank — and shouted: 'Beat you!'

There was turmoil in the kitchen. There was a sense of disappointment for me and my Dad at losing by such a small margin. There was elation for the other team which was, after all, made up of the oldest player and the youngest player in the competition. Avril was especially excited; she jumped up and down and dissolved into a plethora of whoops and cheers. Grandad sat there with a satisfied grin on his face. I will always remember that grin.

### An early lesson in coal mining

Grandad had three greenhouses on his smallholding. There was 'the Big Greenhouse', which was about 40 feet long and 15 feet wide. It was made of wood with glass along one side, the roof and both ends. The other side was a wall to a storage shed and the coke shed. The greenhouse had beds around both sides and along one end and then there was a wide bed along the middle. There was a well under one end of the greenhouse that collected water from the house roof and the greenhouse roof which was used for watering the plants. Its main crop was tomatoes, and when it was fully set it held around 70 plants. There were large diameter heating pipes around the greenhouse that were fed by a boiler to provide heat in the spring-time when the plants were first set.

The other two greenhouses were 'the Little Greenhouse' and 'the Other Greenhouse'. The Little Greenhouse was quite small — no more than 10 feet by 6 feet — but it had a boiler and circulated hot water which meant that this was used to raise plants from seed. The Other Greenhouse was attached to the Little Greenhouse, but it was much bigger — 15 feet by 12 feet — and it was possible to open a valve in the Little Greenhouse and let some hot water circulate around it. Its main role was to grow lettuce and cucumbers as well as tomatoes.

Once the fire was lit in the Little Greenhouse in the early months of the year and the seeds were set, the fire was kept in day and night. It was a cosy place and it was not uncommon for Grandad to spend the last daylight hours there tending his plants, or just sitting and talking to anyone who had come to see him. As you sat you could hear the hiss of the boiler and the hot water circulating through the pipes, and you could almost feel the plants growing.

I recall one night sitting with him as a young lad of about ten years old when he reminisced about his time as a coal miner. He described the coal seams and how the miners had to drive roadways forward to develop the reserves. He explained how they had to use their picks to undercut the seam and then use the cleat lines of the coal to break it down and fill it into the pit tubs. He talked about setting the wooden supports and coping with difficult roof conditions. The talk went on for a long time about many different incidents he recalled. In fact, a lot of the facts and information was way over my head and I couldn't understand it at all, but he presented his story with enthusiasm and pride. He had no regrets. He was assured about the skill and expertise of himself and his mates. The reminiscences went on for so long that it was dark by the time he decided that we should pack up for the night. He regulated the boiler so that it would stay in overnight and filled it with coke. We then went into the house. Grandma was not pleased.

'Where have you been while this time? It's dark outside.'

'We've been in the Greenhouse,' replied Grandad.

'What have you been doing?'

'I've been talking to the lad.'

'What on earth have you been talking about for so long to let it get dark?'

'I've been telling him about my experiences in the pits.'

'You don't want to be filling his head with all that rubbish; he'll never have to get involved with pits and mining.'

Grandma might have been justifiably cross because Grandad had stayed out longer than usual, but she was wrong about the value of that evening. Grandad's father, as well as his grandfather, had been miners and that night I was given an injection of the family history and heritage in the mining industry. It became a part of me. When I chose mining as my career some years later Grandad's recollections stayed with me and I became a part of that heritage.

The motivation in any career can have many sources. For some people their motivation is to be better than their peers; for others it's to earn money and security for themselves and their families, but it is difficult to know when that objective has been achieved; for yet others it is to achieve a position of status or fame. Some are attracted to becoming a person of note in the industry or in their locality.

For me, the challenge was to achieve approval from my forefathers for what I was doing throughout my career in the coal industry. I was always sad that my Grandfather died before I could take him underground and let him see the modern coal mining methods and the equipment involved. I think he would have been impressed by the horsepower that mined the coal instead of the sweat and toil of the miners with picks and shovels. But I was forever thankful for that night in his Little Greenhouse when he passed on to me the baton of his experience, skill and enthusiasm for being a miner.

## *An attack with a pot shovel*

One of the successful initiatives in the Second World War was the 'Dig for Victory' campaign. Schools took on spare land and grew vegetables which were used for school dinners. Many households dug up lawns and flower borders to grow fruit and vegetables. My Grandad also took up the challenge.

Not all Grandad's land was cultivated before the war, so he had an opportunity to bring more land into use. He reduced the free-range area for the hens and this released quite a large piece of land for cultivation. He got the neighbouring farmer to bring in his shire horses and plough up an area of grass. He decided to set it with wheat which could be used to supplement the hen food.

Growing the wheat, harvesting it and threshing it reverted to pre-machinery practices. The seed was sown by hand; the ripe wheat was cut with a scythe; the sheaves were stooped to dry and then the harvest was stored in a shed near the pig sties. The threshing process was the ancient practice of laying the sheaves on a concrete floor and flaying them with sticks to knock the wheat seeds out. The seeds and the chaff were then taken outside and thrown up in the air to let the wind blow the chaff away (from whence came the expression of separating the wheat from the chaff). There was an end product, appreciated by the hens, but the process was not very efficient, and some seeds were left in the straw which was stored for pig bedding.

In time it became clear that the straw was being used as a luxury home to several families of mice, attracted by those remaining wheat seeds. Occasionally one of the mice was caught by the cat but the mouse population was growing exponentially. One afternoon Grandad decided that it was a major problem which had to be tackled. He therefore made a plan, which required the combined forces of me, him and the cat. We would start at one side of the store of straw and turn it over into another place, catching the mice as they ran out into the open. Our weapons would be pot shovels. These were small shovels used to feed fuel into small stoves. They had a handle coupled to the shovel part which was about six inches long and four inches wide. They were ideal for the job, but the aim at the mouse had to be accurate. The cat was left to use its paws and mouth.

The battle began, and it was clear that the mice were withdrawing their army as we advanced. There was the occasional mouse casualty, more often than not the result of a speedy reaction by the cat. As we got deeper into the pile of straw the tension rose and we knew that we were due to have a mouse counter-attack. Suddenly, when we turned over another forkful of straw there were mice running in all directions. The cat caught three — one with each of its front paws and one in its mouth. The cat withdrew from the action until it had disposed of its haul. Grandad and I were banging in all directions with our pot shovels and there were a number of casualties on the battlefield. Then it happened. My Grandad cried out: 'One of them has run up my trouser leg. Come here, lad, and have a look. I think it's fast in the middle of my back.'

He turned his back and in the middle of his waistcoat there was a bulge. The mouse was wedged, and it couldn't get any higher.

'Yes, I think I can see it, Grandad.'

'Well hit it with thee shovel,' he replied.

This was a strange order. I took my shovel and gave it a tap.

'Not like that, give it some hammer,' said Grandad.

I hit it harder, fearing for the damage I might do to my grandfather.

'Hit it harder still. It's still trying to move.'

I gave it a couple more clouts and then it stopped moving. Grandad took his waistcoat off and pulled his shirt out and the mouse dropped out. I think he tied some string round the bottom of his trouser legs before we returned to the battlefield.

When we had turned over all the straw there was a contented cat, who was stuffed with mice, and a further pile of dead bodies.

I realised that I needed to tell my mothers about the afternoon's work.

'You ought to know that I have hit my Grandad,' I said. My mother immediately reacted with shock and concern.

'You've what?' she shouted.

'I hit my Grandad with a pot shovel,' I replied. 'He made me do it,' I added with emphasis. I then told her the whole story, however she didn't take my word for it unconditionally. She went across to next door to see Grandad to make sure he was alright. She found him in the kitchen with Grandma. Grandad assured her I had told the truth and that he was fine.

'He didn't hit it hard enough at first,' he complained.

'Ee, what you fellows get up to,' was Grandma's summary of the afternoon

## *Keeping pigs*

Grandad's pigsties were not big by modern standards, but in the 1930s the building could be described as well constructed and effective. It was about 60 feet long and 15 feet wide. Down the left-hand side was a passageway three feet wide — just big enough to take the wheelbarrow used when cleaning the sties. There were four pigsties — two big ones and two smaller ones between the big sties. Inside each sty there was a trough alongside the passage wall and chutes in the passage allowed food to be poured down into the troughs. The passage stopped at the start of the fourth sty and this therefore had the biggest area. It also had the disadvantage that there was no access to the feeding trough from outside the sty. The buckets of food had to be taken into the sty and poured into the trough, which was a challenge if the pigs were still inside and very hungry.

Just before the fourth sty there was a door that let the pigs out into a sizeable run where they could exercise and dig up the soil with their snouts. It was often the practice to let the pigs out from each sty to have a run out before they were fed. When they got the smell of food, though, they forgot about exercise and banged against the outside door screeching for admission. When the door was opened there was a mad race down the passage and into the sty to get a good place at the trough. As the pigs grew bigger there was much shoving and pushing at the trough to defend the best positions.

It is often suggested that pigs gobble their food, and this might appear to be the case when they are observed feeding, but they are very selective. They do not like turnip or swedes and if there was any of either vegetable in the swill it would all be left in the trough at the end of the feed, even when the trough was licked clean. After the feed, when the pigs were full, they would go and lay down all piled together in their bed of clean straw and go to sleep. After a short while the only sounds would be a chorus of snoring and farting.

Pigs are also very clean in their toilet habits. Their urine and droppings were maintained at one side of the sty so that the sleeping quarters were on clean straw at the other side. Every few days there had to be a clean out and the pig manure of straw and droppings had to be loaded into wheelbarrows and taken onto the muck heap where it naturally heated up and, after a few months, made an ideal fertiliser for the garden.

As one entered the pigsties building the first area was for food storage and preparation. There were bins down the side to hold the different pig meals and corn for the hens. At the far end was a set-pot which was heated by a fire underneath it. This was filled with vegetable and food waste from nearby houses and the local school. At times, small potatoes and other unsold garden produce were used. Usually a set-pot-full was processed each day. When it was cooking there was an appetising smell of a vegetable soup around the smallholding. The product from the set-pot would be quite sloppy. It was mixed with pig meal to make buckets full of pig food. The pigs required feeding twice each day and one of those feeds would likely be hot food. This was a time before the days of bulk feeding with just dry meal and water.

The cycle for pig rearing started when the sow was on heat and had to be served by the boar. There was no fancy artificial insemination in those days. Sometimes the boar was brought to the sow by a local farmer. The sow and boar were let out together into the outside pig run and observed to see if there was the necessary mutual attraction to allow copulation to take place. It didn't always happen as boars and sows can sometimes be picky about their partners. On a few occasions it was necessary to take the sow to a boar. This involved guiding the sow on a walk of over a mile along roads and tracks to a farm by the golf course in Dodworth. This was a tricky operation as there was the risk of dogs upsetting the sow, or of the sow wishing to head off in the wrong direction. Sows do not take kindly to forced marches.

The awesome miracle of birth was always memorable. It often occurred at night and the sow would be laid down on a bed of straw, usually in one of the small sties. There would be a low light burning, so that the process could be observed. Often, as the piglets were born, they would be given a wipe and put to the head of the sow so that she could see that they were alright. They were then put to one of the sow's teats. As the numbers increased and the teats were occupied it became a question of whether she could feed them all. Sometimes it was necessary to raise a couple of the piglets by bottle feeding. Some sows were very careful and knelt down slowly when they were ready to feed, which gave the piglets a chance to avoid being crushed; other sows were casual, and it was a tragedy to lose young piglets, which were fit and well, due to a clumsy sow.

One process that was carried out when the piglets had been separated from the sow was castration of the males. This was a noisy affair, with screams of fear initiated by the smell of blood. The local butcher would come with a very sharp, short knife and carry out the operation while the piglets were held upside down clamped between the legs of an assistant. It didn't last for long, but the screams of the pigs rang around the neighbourhood intimating that something horrible was taking place — which I suppose it was, if you were a small, male pig.

The final stage in the pig process was to send them off to market. There was just room for a lorry to back along the drive and past the greenhouses to the door of the pigsties. The lorry's back door was lowered, forming a ramp for the pigs to go up into the lorry. Side guards were put alongside the ramp and the passage to the sty was cleared. The pigs were then let out and encouraged with shouts and prods from sticks to go up into the lorry. It was not usually a problem with the bacon pigs, which were used to being let out into the run, but it was a different story with old sows, which had a canny idea that something undesirable was being proposed. So the men had a different approach for a sow: they tied a rope around the pig's nose and ran it out along the passage and then over a high side rail on the lorry. Someone was then designated to hang onto the rope and keep it tight as the sow advanced. The others then went behind the sow and shouted and poked its backside and pushed it forward. On one occasion my mother was given the job of hanging onto the rope. She did it very effectively, but this particular sow suddenly reversed, and my mother was hoisted up into the air. She was not best pleased and made her views known when my Dad came home from work.

The daily cooking of the swill in the set-pot carried on throughout the winter and it created a warm corner that attracted Grandad's mates to call in for a chat. There was a short plank that formed a seat near the set-pot fire. One of the regulars on that seat was Wilf Marshall. He was a small fellow, like Grandad, but about ten years younger. One of his claims to fame was that, very late in life, his wife had provided him with a son. The lad was quite bright and passed the eleven plus scholarship to go to the Grammar School, an achievement way beyond the expectations of his parents.

In the days before betting shops and the legalisation for gambling, speculation on the horses was an underground practice. I am not sure whether Grandad placed bets on a regular basis, but on some occasions Wilf arrived with information that he classed as 'A Sure Thing'. On those days the discussion in the pigsty would be animated and there would be an element of tension around. Whether the 'Sure Thing' proved to be a winner was never reported, as far as I can remember. Any investment by Grandad was carried out with great secrecy as Grandma, as a strong Methodist, did not approve of gambling. Whether horse racing had any bearing on the situation I don't know, but Wilf Marshall committed suicide. It was after Grandad had died, and the pigsty was no longer a thriving den of discussion and debate on a wide range of subjects, as well as horse racing.

### A Christmas present

Each year Grandad selected one of his pigs for special treatment. The chosen pig would not be sent to market as a bacon pig but would be retained to be fattened up in luxury. These pigs would become quite tame and when Grandad called them they would jump up and put their front legs on the wall of the sty and Grandad would scratch them behind their ears, which they particularly enjoyed. They had special treats like fallen apples and other titbits from the house. The consequence of all this special treatment was that they grew to a significant size. It was normal for them to be in excess of 25 stones as Christmas approached.

During the war it was necessary to get a licence from the local authority to kill a pig for private use. The licence stated that the pig could only be consumed by the owner and his family, and the person getting the licence had to sign to this effect.

I remember one old lady commenting in the office what she thought of this licence condition.

'We'll eat 'til we bust and then we'll bury the rest,' she informed the officer.

The pig would be killed and cut by the local butcher at the small slaughterhouse he had in the village. By this time Grandad was so attached to the pig that he would take no part in the operation, so this left me and my Dad, with a local relative or two, to get the pig out of the sty, along the drive past the two houses, and into the road. A very large pig does not move easily and it is easily distracted. It was about 200 yards up the road to the slaughterhouse and, although there were very few road vehicles at that time, pigs have no sense of road safely or traffic rules. When we reached the slaughterhouse the pig entered a containing yard which it realised had been used by many other pigs before this occasion. There was a change in mood and the team had to be supplemented to persuade it to enter the slaughterhouse itself. It called for skill and deception, not force. One year a bulky cousin of mine from Doncaster decided to take on the sow and use his strength. The sow swung its head around on his approach and knocked him flying into a heap of pig manure. He had to retreat from the scene — a smelly mess — and go for a change of clothing. The pig was eventually persuaded to enter the slaughterhouse by the rattling of buckets, suggesting food, and other devious ploys. Once the door was shut responsibility was left with the butcher.

The pig reappeared the next day, fully butchered into its varied parts, in a series of washing baskets and containers. All the pig was there, including its head and feet. It seemed amazing that so much meat was contained in that one carcass.

There then began the processing of the different parts. The hams and the sides were salted for several days with saltpetre until they could be hung up, covered with muslin, in the roof in a cool part of the house. The salting was done in the pantry and the walls of the pantry were forever affected by the salt. The large blocks of fat in the carcass were chopped up and heated to provide containers of lard which was always in short supply during the war. The head was cooked, and all the small bits of meat and the cooking liquid were compressed with weights in containers to form brawn which could be sliced as a cold meat. Other parts were made into pork pies and sausage meat. Some of the meat was ground up and made into potted meat, which was a popular spread for sandwiches in those days. The best cuts were roasted as the meat at main meals. For several days the house was dominated by the pig, so much so that my mother began to wonder if it had been a good idea in the first place. But over the year the many, many meals of the meat, along with fried bread and chips made using the lard, proved a great advantage. So, early in the New Year another pig was selected for the special treatment.

While the pig was effectively shared between Grandad's house and my

Dad's house there was a significant surplus and there were no deep freezers in those days to store the meat long-term. Fortunately, as it was war time, there were no streetlights. So, in the late evening, for a few days, there would be a steady flow of visitors emerging quietly out of the darkness. They would always come wearing long coats and when they left they would have bundles under their coats. The ladies looked suddenly pregnant and the men had instantly developed beer bellies. There was no risk of being caught with bags full of newly-slaughtered pig. Thus, the spirit of the licence to kill the pig purely for the family was maintained as far as the outside world was concerned; but some friends did get a welcome Christmas present.

*Air raids*

During the early 1940s, when the UK was subjected to systematic air raids, like the whole community, the Massey family became involved. Firstly, there was the construction of the air raid shelters. Grandad had one, and we had one. They were the Anderson-type, made of shaped corrugated steel sheets coupled together to form an inverted U-shape. They were about five feet wide and of varying length, depending on the number in the household. They were usually sunk into the ground by about three feet and were then covered with soil to disguise them and give them extra protection from bomb damage. If the site became waterlogged the shelter became a minor swimming pool that had to be drained before it could be used. The shelters were equipped with benches along the sides so that it was possible for the family, particularly the children, to sleep through the raid.

The school air raid shelter took the form of an underground tunnel in land adjacent to the school. It was long enough to hold all the scholars — well over 200 — sitting on benches along each side. The Headmistress realised that the successful use of the shelter depended on getting the children into it as quickly as possible once the siren sounded. A plan was devised of where the classes should assemble, and in which order they would enter the shelter. It was the senior classes first, who had to run through to the far end of the shelter. To prove that the plan worked the

Headmistress frequently carried out practices where all the children, each carrying their gas mask, ran into the shelter, class by class, as quickly as possible. She observed the exercise holding her wrist watch to check the time; woe-betide anyone who fell or lagged behind the person in front of them in the gallop into the shelter. It was remarkable how quickly the whole school could be transferred from the classrooms into the shelter; on a good run the whole exercise was completed in well under ten minutes.

My father was in the Auxiliary Fire Service and he had to be on duty some nights each week. The first real fire he attended was in a local sweet shop and he was disappointed to have to push his way through bottles of sweets to get to the fire. When the siren sounded, even if he was not on duty, he still had to get to the fire station and join his mates. He donned his firefighting uniform, which included an impressive belt which held an axe for breaking into buildings. He went on his bicycle to the fire station, which was about a mile-and-a-half away in Darton, but it only took him a few minutes as it was all downhill.

Grandad had his own procedures to carry out when the air raid siren sounded. He had a pile of old hearth rugs and sacking distributed around his smallholding. He went around and soaked these with water so that they could easily be thrown onto any fire to smother it. He also filled up buckets and containers with water from the wells and had them strategically placed so that water was immediately available if any fire was started.

Some bombs were dropped on a coke works within half a mile of our houses, but the main air raids were on the major industries in and around Sheffield. The attacking planes could be heard in the sky and the results of the bombing began to appear as a red glow in the sky to the southwest.

It was common practice for the adults from the local houses to gather together in the road to see what was taking place and to raise their spirits by chatting together. Their discussions ranged from family matters to local village gossip, but also included views on the progress of the war and an assessment of the severity of that night's air raid. On one occasion, as they were talking, the local air raid warden approached them. He was the headmaster of the Higham secondary school who was noted for his eccentricities. He was wearing his long coat, his hat, scarf, gloves and he had both his ears stuffed with cotton wool. He stood facing the group and raised his arm and remonstrated with them for making a noise.

'Hush, they will hear you above,' he said. How on earth an air crew could hear the conversations in a village street in Yorkshire was beyond belief. Even if the crew members could have heard the conversations, and if they had an understanding of the English language, they would have been fooled by the broad Yorkshire accents!

I suppose this is a perfect example of a willing volunteer carrying out his duties without the necessary support of a tiny bit of scientific knowledge to help him. It must often happen in wartime.

## *Good Friday*

Good Friday might be an important date in the church calendar, but it was also the start of Grandad's gardening season. It didn't matter whether Easter occurred early or late in the year; it didn't matter whether the weather was suitable or not; Good Friday was the time to start setting the garden in those parts of Yorkshire. As far as Grandad was concerned he had his plans and he expected support from his family on that day.

Granddad had acquired a Swiss-made, petrol-driven machine to work the garden. It was called a rotavator. It was 3.5 horsepower and had two tractor-type steel wheels at the front to pull it along. It worked the soil with a rotating drum behind the engine which had spring steel tines which cut a track 18 inches wide and to a depth of about seven inches. It had two speeds, that were changed by fitting plugs into the wheels. Low speed was half the high speed and it was possible to dig much deeper in low gear. It was also possible to fit a ridging plough behind the machine to cut a deep furrow. The rotavator had long handles for the driver, which could be used directly behind the machine or swung to either side of the ploughing line. Its controls consisted of a throttle and two clutch handles; one clutch handle started the wheels and the other started the rotating tines.

*Roland driving the rotavator*

The rotavator was a challenging machine to drive, particularly in rough ground, and it was important to keep the wheels level to plough in a straight line. It was too lively for Grandad to drive, so it was usually driven by my Dad or one of his brothers on the Good Friday exercise.

The vegetable area of the garden consisted of two blocks 30 yards long by 25 yards wide. The first job was to set the potatoes. The rotavator would cut out two furrows in the block that was to be set with potatoes. Other helpers would be deployed with wheelbarrows to fill the furrows with pig manure from the stock pile. The seed potatoes would then be set in the furrows from boxes where they had been sprouting. Grandad might do that job to make sure it was done right. There was a real bustle of activity to keep the process moving in line with the preparation of the rows by the rotavator. When several rows had been set the rotavator would drive between the rows and ridge the soil so that the rows of set potatoes were covered in a ridge of fine soil.

By the end of the morning all the potatoes would be set. The family helpers would retire for lunch and to rest their aching limbs, but there was satisfaction to see the large block of garden successfully completed.

After lunch, the rotavator would then go on to prepare the land for other crops; an area for onions and an area for peas and broad beans would be ploughed over. These were usually set by Grandad over the next few days. The rotavator would then be driven into its garage until it was needed for the next area to be ploughed.

There was a sense of relief among the family members at the end of the day that they had been able to answer the call to duty from Grandad. He had a satisfied grin on his face when he went back into the house and spoke to Grandma at the end of the afternoon session. She was not really interested but, as far as Grandad was concerned, she had to be told the good news. It might be Good Friday, but he had a satisfying story to report: he was back in business, his garden had been re-born for another year. It was really Easter Sunday — a resurrection day from the garden's perspective.

### *A pig tragedy*

A smallholder's worst nightmare happened when I was quite small, but it had a bad effect on Grandad. He was very thorough at keeping his stock and his pigsties clean. When they were emptied they were sprayed out before new stock was introduced. The risk of disease should therefore have been negligible.

However, a time came when the pigs appeared to be ill and the vet was asked to investigate. It was Swine Fever. All the stock would have to be destroyed. I still have visions of my Dad, in the kitchen, telling my Mother, and her face reflecting her shock and despair.

'What effect will it have on Grandad?' she asked. She was very close to Grandad and he often came and talked to her rather than discussing things with Grandma.

'He will be devastated,' replied my Dad. 'I suppose the procedure will be laid down and supervised by the Ministry of Agriculture.'

A large hole about six feet deep was dug in the pig run. The pigs were shot and then they were all tipped into the hole. The big sows were mixed in with the smaller pigs in a massed grave. The bodies were then set on fire and the smell of burning flesh could be detected in the surrounding houses. The grave was then filled in and the area was covered with large stones to avoid it ever being dug up.

I suppose that my Dad did most of this work as Grandad could not bear to deal with this treatment of his animals.

All the buildings were sterilised, and, after a time gap, pigs came back onto the scene. I suppose Grandad regarded it as just another tragedy in his life; this time it was animals, not his children.

## A family sing-song

Grandad's house had an exceptionally large front room. It had a large table dominating the room, with limited furniture around the edge — a horsehair sofa and the odd chair. The family organ dominated one wall. It was possible to squeeze 20 people around the table during family get-togethers and on some occasions even this was not enough, and the children had to sit on an additional table in the hall. Fortunately, the family did not live at great distances from Grandad's house in Higham, near Barnsley, so get-togethers were not difficult to organise. Harold lived in Goole and Charles lived on the outskirts of Doncaster and they had the furthest to travel; Olive also lived in Doncaster, in a licensed beer-off in Bentley Road before moving to the next village, Redbrook, when they retired; Annie lived in West Melton; Albert lived in Higham and Roland lived next door; Ellen was still living at home with Grandma and Grandad. If they were all there with the grand-children, the total would be 23.

Feeding such a number did not seem to be a major challenge. There was always lots of home-baked bread. There would be ham from the last pig that was killed at Christmas. Charles's wife, Lucy, had a shop and she could be relied upon to bring tinned food from her stock. There were always eggs from the hens. If it was summer-time there would be tomatoes and lettuce from the garden and there would be strawberries and raspberries. There would then be freshly-baked buns and cakes. While the men wandered around the garden looking at all the vegetables and fruit, the ladies would be in the kitchen cutting bread and preparing the food for the tea. There was also a team setting the table. Grandma had relatives who were in the crockery business in Cheshire, so there was no shortage of high quality crockery. Someone else would have the job of sorting out the chairs and forms to seat everybody. There would be two kettles on the open fire in the kitchen to fill several large teapots. After the first mashing of tea was served there would be more kettles of boiling water to allow for second cups.

Grandma and Grandad always sat together in the same place on one side of the table, with their back to the bay window. There was no real pecking order for the rest of the family, but Olive, as the oldest child, if she was there, assumed the responsibility of seeing that everything passed off alright without any worries for Grandma. The conversation during the meal was about family matters and the children were instructed to keep quiet and use their best table manners.

When the meal was finished the table would be cleared and all the food and crockery taken through into the kitchen. Any left-over food would be packed away in the pantry and a production line would be deployed to do the washing up and put the crockery away into the special cupboards in the front room.

Grandma and Grandad stayed sitting at the table as all the activity carried on around them. Grandma made the next move.

'Are you going to give us a tune on the organ, our Dad?' she said to Grandad.

If he agreed, then the organ stool would be found and he would move to the other side of the room. He would search out some appropriate music book. He had a range of hymn books including the old Moody and Sankey tune book.

The organ was made by DW Carn and Co of Woodstock, Canada and was marketed by Robert Stather of 197 Seven Sisters Road, Finsbury Park, London. It was a single manual, 14-stop organ. It had foot pedals to pump the air and two knee swells to increase the volume. Grandad had never had any lessons but had taught himself to play. He used the stops on the organ to vary the sound according to the words of the hymns. When he wanted a fuller sound and volume he used the knee swells. As well as the organ he played violins that he repaired and re-strung.

Once Grandad started on the organ, the activities in the kitchen would be bought to a rapid conclusion and the family would gather behind Grandad to sing hymns. My mother, Vera, would lead the sopranos; Olive and Annie sang contralto; Roland and Albert sang tenor; Charles and Harold sang bass. It was a magnificent sound of balanced harmony. Grandad had a smile on his face as he played; he loved good music. There were requests for special hymns, that he was glad to play if he could find them. The music had an aura about it that everyone there recognised was not experienced by many other families. Some of the grandchildren joined in to be a part of a unique occasion.

Grandma would stay in her place at the table, but she would sing the hymns as she knew all the words. She often sang with one arm on the table holding her head and with her eyes closed. The ladies kept an eye on her, but no one disturbed her solitude. Was she revelling in the hymn singing as an expression of her faith? Or was she singing and praying at the same time, thanking the Good Lord for the joy of having her family around her?

### *A faithful dog*

After the death of Judy, my Grandad's dog, he got another one which was called Rex. Rex came as a puppy but, as Grandad was getting older, the dog spent most of his time with me. He was a Labrador/retriever and his speciality was chasing balls and fetching them back to the thrower. He was so good at finding balls that the Higham cricket team hired him to search in the corn field next to their cricket pitch to retrieve those knocked out of the ground with big hits. He was also a good excuse for me to take a walk into the local woods if a girlfriend was also taking her dog for a walk.

When I was a teenager Rex was an essential part of a form of cricket that two of us designed. One of us bowled and the other one batted. Rex did all the fielding and he was trained to deliver the ball back to the bowler after each hit. We played in the grass field that had been ploughed up to grow crops during the war. It was back to its role as a free-range hen run and it was bounded by hedges. Rex had the skill to get through the hedges and retrieve the ball if the batter succeeded in making a big hit. On one hot afternoon he ran until his tongue was hanging out and he sat on the boundary panting furiously. I believe that was the first cricket match stopped to revive a dog with a pail of water.

Rex's most important job was to guard me on the first leg of the journey to Wakefield when I started at the Queen Elizabeth Grammar School in September 1945. It was a long journey for the three of us from Higham — John Marshall, Derek Hinchliffe and myself. We walked to Barugh Green, about half a mile, and then caught a bus to Darton; we walked through the village of Darton to the railway station and then took the train to Kirkgate station in Wakefield; it was then a twenty-minute walk through Wakefield to the school. Our journey started at 7.30am and Rex joined us in the walk down to Barugh Green. When the bus arrived, we threw a stone for the dog, which he collected and took home, depositing it on the step to tell my mother that we had caught the bus.

# Chapter 4

*A new world,*
*Queen Elizabeth Grammar School Wakefield*
*1945-1951*

Queen Elizabeth Grammar School Wakefield was founded by Royal Charter in 1591. It moved to its current site in Northgate, in 1854. After the decision was made that I should accept the George Beaumont Scholarship and go to the Grammar School in Wakefield, my parents visited the school to be briefed on its objectives and standards before I started. They were advised where to buy the school uniform and sports kit, all of which was mandatory.

The three of us from Barugh Green School travelled to Wakefield for the afternoon session on our first day in good time so that we were able to get an early look at the school. We were briefed by an older boy in the playing fields behind the school. He painted a picture of strict masters and tough studies, and also plenty of homework. He laced his comments with stories about the boarding house, some prefects who had maniacal tendencies and the use of detentions to guarantee discipline. His pep talk did nothing to calm our nervous fears, and the playing fields, which had rugby posts, were further evidence of major change for us who came from soccer country.

In the afternoon, at our first school assembly, we got an impression of the size of the school and its imposing array of teachers — nearly all men — dressed in gowns displaying the colourful sashes of their degrees. The assembly was taken by the headmaster, W A Grace, whose nickname was 'Nodder', and he also carried his mortarboard cap to the lectern. Listening to the language from the stage, and the speech of most of our fellow students, we realised that our broad Yorkshire accents would have to be modified if we were not to stand out.

Our form was 3A. Our form master, W V Vincent, gave us a full briefing of how the school worked and he gave us our timetable. He also gave each of us a Record Book. This was divided into a small space for each lesson and good or bad performances were noted by the initial of the master concerned. This was to be a continuous measure of your position in the league table of students in the class. At the end of the first day the three of us knew that, from being the stars of our junior school, we would be challenged to achieve any distinction in this form. Within days, one lad called Watson was given a bad record. I was amazed to see him eating his lunch in the refectory as though nothing tragic had happened. A few days later I got my first bad record, from Miss Denton the Latin teacher, for poor marks in a vocabulary test that she had dropped on us. I was distraught when I arrived home. My mother took a positive approach and waded in to coach me for the retest. I shall always remember that 'rogare' is the Latin verb for 'to ask'. Mother said, 'Think of robbers — they are rogues and they never ask.'

The school had one distinguishing feature compared to other schools — it was open for classes on Saturday mornings. This made it a long week, as we left home at 7.30am and didn't arrive home until about 5.15pm each day. Tuesday and Thursday afternoons were set aside for sports. Rugby did not attract me, nor did cross-country running. I had hopes at cricket, but I was rejected after a short two-over trial for the under-14 team. I was told that because I played back rather than forward to the bowling I had no future as a cricketer. So, on some Tuesday and Thursday afternoons I was able to return home early.

While my school reports displayed a mediocre performance there was an additional element of our education in the journeys to and from school. On the trains there were other students, both boys and girls, from Barnsley,

Haigh and Crigglestone, as well as Darton. We were able to mix with them in a sort of club, sharing each other's experiences of our privileged educational trials. Of particular note was Ronald Eyre, who often travelled in our carriage from Darton. Ronald later became a television performer and producer of some note. He had an infectious sense of humour, which was often evident on our journeys. He walked with very splayed feet and he often mentioned that my feet were both perfectly straight as I walked. He was a Methodist with firm beliefs derived from his family's influence and his position was similar to mine. I took comfort from him in my early years in the challenging environment of the school.

The school had a classical tradition, so we studied Latin and Greek as well as French. We also studied Maths, English, History and Geography and Religious Instruction. On the science side there was General Science and it was only with clear specialisation that one took Chemistry and Physics courses. We were locked into a four-year cycle to take the School Certificate examination. The sequence of classes that I followed to School Certificate was 3A, 4A, 5B and then 5UpperA. The year in 5B was a clear indication that I was not rated in the fast track stream. Being reinstated into 5UA might reflect that I had made better progress during the third year.

# Chapter 5

*Major changes at home;
Farewell, Grandad*

Grandad was dying. It was not talked about, and I was never told directly, but I knew. He was confined to his bed and my mother spent a lot of time trying to make him as comfortable as she could. Grandma knew what was happening, but she was lame and couldn't help him. Grandad's youngest daughter, Helen, who was still at home, didn't have the constitution to cope, and sat in the kitchen and cried. So, it was left to my mother to nurse her father-in-law. The doctor came to see him but there was little that he could do as it was stomach cancer.

I recall at this time that a lorryload of small coke was delivered and tipped outside the coke shed alongside the Big Greenhouse. It was a large heap and it had to be thrown into the shed. I made a start and tried to make as little noise as possible so as not to disturb Grandad. During Grandad's illness, Dad got involved in the smallholding to keep it going to the best of his ability, as well as holding down his full-time job of repairs to the colliery houses. He came along when he came home from work and helped me complete loading the coke into its shed.

Grandad loved music, and an opportunity arose near the end of his life to let him hear a special musical event. There was to be a broadcast of the annual performance of Handel's Messiah, given by the Huddersfield Choral Society. The soprano soloist was Isobel Baillie, whom Grandad admired greatly. I am not sure whether it was Grandad who requested it, or my Dad who suggested it, but it was decided to establish a radio in the bedroom. This was complicated, as radios, or should I say wirelesses, in those days were not designed to be moved around.

After significant trials the radio was persuaded to work and the wonderful sound of 'Comfort Ye, Comfort Ye, my People,' was clearly heard in the bedroom. If anyone needed comfort that night, it was Grandad. He lay in the bed with only his face showing, but the bed clothes were pushed up by the swelling of his stomach. His white hair was parted and neat, but his face was grey and his eyes, when he opened them, showed his awareness of the end and the agony of his helplessness.

Grandad listened to the music, often with his eyes closed, and at times it was difficult to know whether he was listening or sleeping. We kept a vigil of silence, not knowing what to do for the best. Did Handel realise that his oratorio would be used to support a soul on its final steps to death and the great unknown? The oratorio continued into its second part and we wondered if Grandad's attention would stay until Isobel Baillie sang his favourite aria. He had a preview in Part Three when she sang 'How beautiful are the feet of them that preach the gospel of peace', but there were more choruses and solos until the dramatic 'Hallelujah Chorus' rang out. Then we were there. The soprano aria, 'I Know that My Redeemer Liveth'.

The simple, moving aria of faith was proclaimed by the bell-like voice of Isobel Baillie. Grandad opened his eyes and listened. In the aria the words were repeated many times as the music was developed. 'I know that my Redeemer liveth and that He shall stand at the latter day upon the earth. And though worms destroy this body, in my flesh shall I see God.' Then there was a new musical theme and the words were acclaimed dramatically and forcefully, and repeated several times, 'For now, is Christ risen from the dead.' In the final bars, very quietly, the last phrase is veiled in mystery but also hope, 'The first fruit of them that sleep.'

The words of the aria might not have moved Grandad, as he was not a church-goer like his wife, but through the music, for a few minutes, that bedroom was converted into a place of sublime peacefulness, a waiting room for paradise. Grandad indicated that he had had enough, and the radio was switched off. He closed his eyes and settled down to wait for death. He passed away on the second of December 1946.

### Poor Grandma

When Grandad's will was read, it revealed a healthy position. There were no debts and there was over £1,400 in the bank. His market gardening had been successful during the war and his careful use of his land had paid off, no doubt helped by a few special tricks he had. One I recall applied to the cabbages he grew: he would cut the cabbage when it was ready, but he would leave the stump and root in the garden; in time, three or four small cabbages would burst out of the stump. He would also cut them and sell them when they were ready. He grew sprouts successfully. I explained to him that the sprouts we had tried to grow in the school garden during the war did not form tight sprouts but just a few open leaves. He explained that it was essential to keep the ground around the sprout plants very firm and solid — if necessary by trampling on it.

Grandad had explained that he wanted his youngest daughter, Helen, to benefit from his will. She was still at home, and she was essential as support for Grandma. Helen was not retarded but she was simple-minded and had a strong character. Much to my mother's irritation she could often not be relied upon to do things in the right way or at the right time. Grandma had problems with her legs failing, as they were permanently crossed. She could only move along by using a tall stool (a forerunner of today's Zimmer frame) that she moved forward and then leaned on and pulled her legs forward.

The view was clear to the Massey family that Grandma could not be expected to survive long after the death of Grandad. So, my Dad made an important decision. With the reluctant agreement of my mother, he decided that he would work the smallholding as well as his job. The commercial arrangement was that he would hand over half the profits to Grandma from the operation each year. What was a temporary arrangement became a long term challenge, as Grandma lived a further 23 years, to the age of 94, after Grandad died.

Dad's commitment affected all of us as there were times when essential work had to be done in the garden. I remember one occasion when I wanted to join my mates playing cricket. I was told that I had to pick two rows of strawberries before I could go out. I skipped over them and came in with a small bowl of fruit. I announced my departure but when Dad saw the amount of fruit I was sent back to repeat the exercise. 'I know that there are more ripened strawberries on those two rows than you have picked; pick those rows again.'

Looking back to those years after Grandad died I am amazed at the output of my mother and father. As well as the smallholding and his job, Dad was Sunday School Superintendent at Higham Methodist church and at one stage he was involved in a Youth Club that did woodwork and other productive hobbies. Both Mother and Dad also sang in the choir. How did they do it? They worked consistently hard, seven days each week. I was expected to play my part, but my schooling was always allowed to have top priority.

## *Special church events*

The church was an essential element of Massey family life. It took up all day on Sunday, as there were morning and evening services, as well as Sunday School in the afternoon. The annual church calendar was also peppered with special events.

At Christmas, as well as the carol service, the choir, with helpers, toured the village on Christmas Eve singing carols. They started at 11pm and finished about 3am on Christmas morning, after having refreshments in a house at the half-way stage. The sound of that music was initially in the distance but increased in volume as the singers got nearer. As they assembled outside our house the initial sound was of footsteps and muffled talking as the pieces to be sung were identified by the conductor. At the later stages of the tour they often sang a Christmas morning hymn, 'Hail Smiling Morn.' The tune included lots of runs by the various sections of the choir and ended with a dramatic climax. The singing in the open air was a magical sound, whether one was a believer or not. It truly announced the anniversary of the most significant event to affect the human race.

Then there was the Whitsuntide parade and sing. A brass band led the combined Methodist Churches of Higham, Barugh Green and Barugh, with their banners, through the three villages, with stops along the way for hymn singing. After the parade all the three congregations returned to the Higham Church schoolroom for tea. Three long tables were laid out along the hall and sandwiches, mainly potted meat, were served first. There was often competition between some of the young boys to see who could consume the most sandwiches. After the sandwiches, plates of buns were placed along the table. These plates were monitored by lady supervisors with eagle eyes to ensure that each person had only one bun.

After the food, there were games in a field alongside the church and the three churches competed to see which one could win the most events. The weather on Whit Monday could not be guaranteed, so an alternative competition was arranged in the schoolroom when it was a wet day. Each of the churches manned one of the long rows of tables and answered questions or played games to gain points.

One game that was very popular was a sort of forfeits game. The question master called out the name of an article and the team that delivered it to him first got the point. It could be, for example, a diary, a penny coin, a comb, a handkerchief, a tie, a sock, a shoe etc. This led to members of the teams loosening clothing and getting partially undressed in anticipation of the call for a particular item. When this game was in the programme the question master usually tipped off one person that he would be requesting a pair of trousers. Each time this happened there were always whoops of excited cheers from the ladies; their expectations were not usually fulfilled, as one man dropped off his trousers only to show that he had another pair on underneath.

However, on one occasion the local butcher's son, in a burst of reckless enthusiasm, dropped his trousers and ran out to the front with his shirt flying behind him and he wasn't wearing underpants! He won the point but the return journey to his seat was more cautious as he tucked his shirt between his legs to regain his modesty. The cheering went on for minutes and he was a hero.

There were several special religious events and services in any year, but the most significant was probably the acceptance into membership of the Methodist Church of the young people who had agreed to go forward. They were taken through a series of meetings and teaching sessions by the minister and senior church members and then presented to the church members at a special service. In my year there were about eight of us as a group accepted at the same time and most of us were children of practising members of the church.

We took communion as a group, but I have no recollection of there being any blinding flashes of light in the service or any change in my feelings about my faith. It was, though, an important step along the way. In retrospect, it joins other experiences over a lifetime of moments when I was touched by the Holy Spirit, supported and uplifted, guided forward with added strength and power more than my own. Within the youth members of the church I was active in social events and I took part in Youth Services where we spoke of our religious views. Some people thought that my career calling might be a life in the church ministry.

# Chapter 6

*Further education and two big decisions*

The School Certificate in 1949 was the major event for Queen Elizabeth Grammar School students in the Upper Fifth forms. It would demonstrate their academic ability and be the guide for their studies in the sixth form. I took nine subjects and achieved credit level in five (English Literature, History, Geography, Scripture and French), and pass level in four (English Language, Greek, Latin, and General Science).

The outcome of deliberations on these results was to enter the sixth form to study History, English Literature and Scripture. The potential aim was to study History at university, but there was a problem: I needed a credit in General Science and Latin to gain entry to the better universities. So, my sixth form studies were complicated by having to undertake further work in Latin and General Science.

The studies in English Literature and History gave me a lifelong interest in reading and writing. The teaching of Bill Teasdale, for English, and Ronald Chapman, for History, was truly inspirational. In the sixth form I sat next to David Storey, who became a famous and prolific author after writing 'This Sporting Life', about his experiences as a Rugby League player.

My studies in the first year of sixth form were affected by illness and I lost half a term's work. Eventually I was hospitalised and underwent a serious operation. My appendix was removed but the surgeon found that my colon was 'floating'. He tied it back in place with cat gut and his notes stated that, 'he hoped that it would stay in place.' His hopes were well-founded and over the next seventy years there has been no recurrence of the symptoms or the pain.

There was disappointment at the end of the first year in the sixth form when I repeated the Latin and General Science School Certificate exams. Again, I failed to achieve the required credit rating. On getting the results I recall being in total despair at home and thought that my future plans were completely destroyed. I decided in the second year of the sixth form to concentrate on History and English Literature in the new A-level examination. This was a new exam for both students and teachers and was something of an unknown challenge. But 1950 brought with it two life-changing experiences aside from this.

It was Whitsuntide Sunday in 1950, and it was a special day in Higham Methodist Church — the Sunday School was celebrating its anniversary. A tiered staging had been built to hold the scholars and they had been trained to sing new hymns and recite appropriate poems. They would look very smart, as it was the tradition to buy new dresses for the girls and new jackets, trousers, shirts and ties for the boys. These new clothes had a protection order on them, as they were expected to last for a full year. It was guaranteed that the church would be full, at both the afternoon and evening services, as parents and grandparents came to watch their offspring perform.

The preacher on these occasions was a very special appointment and they were selected and booked well in advance. Whether male or female they had to oversee and coordinate the whole event with suitable words of encouragement to those children who would be nervous about performing to a full congregation. In a busy programme for the service, the preacher's talk to the children had to be interesting and memorable and the sermon to the adults had to be short, with a relevant Christian message. It was a challenging engagement.

On this occasion it was a local lay preacher called Ernest Andrews. He had done the job before and there was an atmosphere of excitement to hear him. He had a magical ability as a storyteller that captured the attention of the children as well as the adults. He always told stories about two young children called Arthur and Olga, who had adventures in their village but always ended up as winners.

It was the tradition that the preacher should be entertained to tea by some church member to fill the time between the afternoon and the evening services. So, there he was being shown around the smallholding that my father ran as well as his normal job, while my mother finalised the tea. We were looking out over an expanse of fruit and vegetables. I was standing at the side of my father, a teenager studying at the Grammar School in Wakefield, when Ernest Andrews asked a question.

'Now then, Roland, what is this lad of yours going to do in life?'

My father had the job of answering the question. He hesitated and then said; 'Well, he is thinking about entering the ministry.'

This was vaguely true, as I was in the first flush of enthusiasm after becoming a member of the church.

Mr Andrews looked at me and shook his head. 'Oh, you have to be absolutely sure that you have the calling before you enter for the ministry.' It was a salutary message that rang true to both me and my father. The Good Lord had sent the message for us to think again.

The second significant event occurred during the Christmas holidays in 1950. The deal was that I had to arrange a visit to a coal mine. I was able to do this because my auntie's husband, Tommy Howarth, worked on a coal face at Woolley Colliery and he was able to fix up the visit on a Saturday morning. There was a shortage of coal at that time, so the pit was in production that day. We met the colliery manager on the surface before we went underground. The party was the three lads, my Dad with Uncle Tommy, and another guide. We saw the coal face in the Parkgate seam where my uncle worked. It was over six feet thick and we passed through the coal face and saw the men filling the coal onto conveyor belts and setting the wooden supports. We were able to follow the coal on the conveyor belts to a loading point where the coal was loaded into half-ton pit tubs. The tubs were hauled to the pit bottom by a sequence of rope haulages and then loaded into cages and hauled up the shafts to the surface.

*6 Modern 2 with Bill Teasdale, our English master*

This photograph shows Bill Teasdale, the English Master, and his class of students in 6 Modern 2 form. He was a brilliant teacher who inspired all his students to study literature and to express themselves in writing. Just behind Bill Teasdale, on his right and winking at the camera, is David Storey. He wrote 'This Sporting life' based on his experience as a Rugby League player. It was made into a film and David had a prolific career as a writer and an artist.

On the back row, third from the left, is Michael Hancock, and fifth from the left is Michael Cooper. I am in front of Michael Cooper and peeping around the person in front of him. What the three of us had in common was that we had no clear career plans. Michael Cooper arranged for us to visit a woollen mill in Huddersfield, where his father was a textile designer. We had a good visit to the mill, but it didn't influence any of us career-wise, though I recall us going to a social event at a local church and dancing with a girl called Tina. She was an amazing dancer; it was like holding a feather in my arms.

We were fortunate to see the whole process of getting the coal from the coal face to the surface. I was fascinated by what we had seen and decided on a mining career. This was a major shock and concern to my parents, as the aim of any parent in those days was to educate their children so that they did not have to go into the mining industry. They thought that they had achieved the objective to give me a better career than coal mining.

I was assisted in making the decision to enter coal mining by getting advice from a senior mining engineer in Yorkshire, Mr H J Atkinson. His hobby was to play the double bass in an amateur orchestra conducted by a personal friend of his, who later became my father-in-law. H J Atkinson was distinctive when driving to rehearsals, with the double bass sticking up out of the boot of his sports car. I was introduced to him and, in the Easter holidays, he arranged for me to have several days looking at other mines under his control to confirm my interest. He then said that I must aim to go to university to get a degree in mining engineering.

I stayed on at the Grammar school and successfully took the A-level exams in History and English Literature. The challenge of these exams was that they were the first exams of the new A-level syllabus, a new format. After taking the exams I left to start a lifetime adventure, gaining the necessary A-levels to take an engineering degree at university, which would present me with a major challenge.

# Chapter 7

*A miner starting at the bottom*
*1951-1954*

Starting work at the pit could not have been a bigger contrast to being in the Sixth Form at Grammar School. I started work on the surface at North Gawber colliery in Mapplewell, travelling to the mine on a push bike and coming home in my pit dirt, as there were no spare lockers in the pit head baths. When I arrived home on my first day the local junior school headmaster was waiting at the bus stop outside our drive. I spoke to him in my pit dirt, summing up how I felt right then: 'Well, that's the end on my first shift at work. I wonder where I will be when I work my last shift?' He looked at me in amazement, but neither of us could have projected what would happen over the next 41 years.

Initially, after six weeks on the pit top, it would be back to school. Because I was under 18 years old I had to join the basic training course which lasted about 12 weeks. We did elementary classes in maths and English language, with instruction on pit operations. One element that was very helpful was a first aid course that taught us how to deal with accidents underground. We were tested on diagnosing what was wrong with patients and what we would do to treat them. In the first aid test I remember that my case was a fractured femur which I successfully diagnosed. I had to immobilise the patient with splints, ready for transport out of the mine. The examiner was impressed by my treatment of the patient. The course also included keep fit exercises in a gym, and there it became clear that some of the young lads were physically very strong and would make excellent workmen. I think at that stage I was more worried about my physical strength to hold down a job in the industry, rather than the technical and academic challenges.

The classes took place on the surface at Woolley Colliery, but we also went underground to a training centre where we were able to practise handling pit tubs on haulage systems and the use of pit ponies. About once each week we were taken to one of the working coal faces at the mine to see various face operations. These visits whetted our appetite to get onto the coal face and start producing coal. In fact, at that time, there was a long delay before coal face training was available and the end of the course usually meant spending several years on work away from the coal faces, manning the haulage systems and servicing the coal faces.

One day we came out of the mine from an underground visit and saw, in the pit yard, industrial relations in the raw. About 30 men were surrounding a union official and they were arguing about a wage issue. We were told that the union official was Elijah Benn, the National Union of Mineworkers Treasurer at Woolley Colliery. He was a dominant person, well over six feet tall, so his head was higher than most of the workmen. They were threatening to stay on the pit top until their case was settled; he was insisting that they went to their work and he would see the manager about their problem. In the end he prevailed, and the men went underground. My thought, at the time, was one of abject fear. How could I possibly deal with such a powerful individual? If I made progress in management I would have to deal with trade union officials; my skills base needed to expand into new fields, I realised.

There was a passing-out ceremony at the end of these courses. The certificate presented was the most impressive I have ever received. With a background of the National Coal Board flag and several signatures by NCB management, it is a larger and more impressive certificate than either A-level results, a university degree or the mine manager's certificate. I still retain it!

The biggest challenge facing me was to obtain additional A-levels in Chemistry, Physics and Mathematics. But I had never had any success in science subjects at the Wakefield Grammar School. How could I possibly achieve that objective? The route lay in attending day release and evening classes at Barnsley Technical College while gaining underground mining experience at the colliery. The secret weapon that the Technical College used to help the course members, was to get the best teachers in the various subjects from the Grammar Schools in the district. The teachers were getting additional pay to add to their other salaries, but all their skills had to be focussed on getting an understanding of the A-level syllabus through to us in a matter of months. In retrospect they were the best teachers of my whole academic career.

I did one year in courses aimed to fill in basic knowledge of science subjects before starting on the A-level courses. However, there were still many occasions when I got stuck in answering a question where I was certain that a piece of information from the O-level syllabus was the missing link. This fear persisted until I reached the university Mining courses.

It took two further years to gain the required A-level grades for university. It would be fair to rank those years of underground work, with day release and evening courses, as the most difficult of my career.

### *Working underground*

After completing the training course, my initial underground jobs were away from the coal faces. I worked at North Gawber Colliery in the Lidgett seam, with a team of belt men who maintained the conveyor belts in that part of the pit. Once, I made a comment about the management of one particular job, at which point a member of the team said that he was the father of the colliery manager. More tact required! Sometime later I was placed by the colliery manager with one other person to carry out special jobs for him. The main one was to measure up the work done in each roadway of the mine at the end of the week. This involved going to the pit at lunchtime on Friday morning and staying through to the end of the

Saturday dayshift and during that time measuring every roadway in the three seams at the pit. This job was not without its challenges. The measurements were taken by the deputy on each coal face and it was not unusual, initially, to find that their results showed an excess over our readings. As the men were paid according to the cubic feet of strata extracted, there was money involved. It needed discussion and re-measuring to show the true figures. It was known, though, that we had been sent by the colliery manager!

On one occasion we were on a new coal face to take some readings, but we couldn't get to do the work until the coal had been cleared. We were helping to clear the coal when the colliery manager came onto the coal face. He realised that, though I was showing initiative, I had not done my coal face training, so I should not have been there. This led to me being moved to the coal face training programme, again in the Lidgett seam. One good feature of the mining industry was that everyone — yes everyone, whatever their qualifications — had to be face-trained before they could be in any official position underground or work unsupervised on a coal face.

I was initially put with a workman who had some exceptional characteristics. He was in his early fifties, he sang in the male voice choir at the pit, he was a beer drinker who seemed to have contacts providing after-hours facilities such that he had but a few hours' sleep each night, but he appeared every morning on time and completed his stint of work with no problems. He had the help of me as trainee, but I recall on some occasions we filled additional coal as well as his stint.

When that period was complete my next assignment was to join a team of men driving the main roadway on a coal face. This was a three-man team where there should really have been four. So, the trainee had to get stuck in! It was heavy work and the dirt extracted was all packed away along the coal face where the coal had been extracted. The roadway was twelve feet wide and the thickness of dirt was eight feet and each day the roadway was advanced by five feet. When the dirt was fired down it was a big heap. When we had moved all the dirt and set the supports we all had aching limbs. Each Friday — pay day — I was slipped some additional cash by the team leader to recognise my efforts, as I was only on a fixed rate of pay as a coal face trainee.

My final job was to work with the machine men who undercut the coal seam. The machine had a five-foot-long jib along which ran a chain carrying the cutter picks. The task was to cut about 120 yards, which was half the coalface. Before the undercutting started the machine had to be manoeuvred into position, any blunt picks changed, and the jib swung around into the cutting position. As the machine went along, the undercuttings had to be cleared away and the wooden supports re-set. At the end of the run the machine had to be stabled clear of the conveyor which ran through the coalface when the men were filling the coal. I was placed with a team that had a reputation of being 'cowboys,' but I got on well with them and I became quite proficient at driving the machine. It was necessary to keep the machine balanced at the right level so that the cut was just at the base of the coal seam.

One exceptional incident occurred during this period. Having finished our shift, we were walking along the main roadway to the man-riding train when we were attracted by an official waving his light and shouting for us to help him. We rushed back and found that a workman on his district had died. The man had been packing the dirt in a roadway where the conditions were very hot. He had come out for a drink of water and had collapsed and passed away. We carried him out of the pit. When I got home I told my father his name and found that he had gone to school with the man. It subsequently became known that he had been advised not to work on the coalface, but found the reduction in his wages too much, so he had returned to face work. It was implied that he had been influenced by his wife in this decision.

### A special coal face experience

On one occasion I was working on a coal face as part of a power loader team. The machine loaded the coal along over 100 yards of the coal face which had been undercut and broken down with explosives. The four-man team had to set the wooden supports and clean up the coalface as well as driving the machine. The conveyor belt had stopped, and nothing could be done until it started again, so we all crouched down to rest. I don't know whether I was whistling a hymn tune or humming some song or other, I felt a tap on my shoulder. I turned around to look into a black and sweaty face. 'Are you satisfied with what the Lord Jesus can do for you?' I knew that the questioner was an evangelical Christian. To that question, in such a strange place, I don't know what I replied, but it should have been, 'Yes, indeed I am very satisfied.'

# Chapter 8

*A new henhouse and a new church*

My father had bought a piece of land adjacent to my grandfather's land and he kept it for several years without developing it. It was during my second year at university (see below) that he decided to construct a large building to raise hens and pigs. As the total landholding of the smallholding exceeded the minimum size to be classed as a farm, he didn't need any planning permission. I offered the expertise gained from a civil engineering course taken that year as part of my degree studies to help in the design, which grew in size as we went along. The specification was a two-storey building with concrete block walls and a roof of curved asbestos sheets. The width was 19 feet and there was a downstairs bay and an upstairs bay at each end, each measuring 45 feet by 19 feet. In the centre was the stairs and storage space at both floor levels for all the feedstuffs.

My Dad was a collector of resources and he had acquired a large stock of wood from an old manor house that was being pulled down because of subsidence. This timber was used for the upstairs floors which were then covered with sheets of asbestos to carry the deep litter which would form the floor for the hens.

Two interesting features affected the building. As the ground was clay I insisted that we should put piles down to the bed rock to ensure stability. These piles were about four feet deep, so the amount of concrete for the piles and foundations was substantial. Large lorryloads of aggregate were tipped up and all the concrete was mixed by hand, so the project kept us busy through the summer.

When the building was complete, and set up with nest boxes and feeding hoppers, a hundred laying hens were introduced into each of the two upstairs sections. The hens seemed to enjoy their facilities and made suitable choruses of clucking and laid large numbers of eggs. These had all to be graded and cleaned and stocked in boxes to be collected each week by the Yorkshire Egg Producers Company. The cheques received each month might have been substantial, but the profits were only generated by the free labour provided my Mother and Dad.

There was one technical problem that had to be resolved. The weight of the deep litter, as it became thicker, caused the upstairs floors to sag, as the 19-foot-long wooden spars were not strong enough to take the weight. I was able solve this problem by borrowing a hydraulic prop from the pit and lifting the floor up to its correct level position. This allowed additional wooden props to be set onto the concrete floor to support the wooden beams of the upstairs floor.

The building was never used for pigs on the lower level as it had been designed; instead more hens were housed in the building and some at the lower level were able to roam outside the building as free-range hens. The boxes of eggs for Yorkshire Egg Producers increased in number.

## Back to music

As a child I had been the unwilling student of a patient piano teacher, who succeeded in getting some basic skills into me, but not to any measurable level. I regarded it as an achievement when my mother decided to call it a day and suspend lessons. However, during the intensive studies for A-levels, as well as working in the pit, I went back to playing the piano and began to make progress. In the end, I recall a few times playing the Radetzky March without watching the score, as my fingers were sweeping over the notes carried along by the spirit of the music.

It was during this time that the Methodist Church at Higham installed a new pipe organ in the church. I decided that it might offer me an opportunity to take my music to a better level by getting trained as an organist. I wrote to the Church Council and sought their permission to practice on the organ. They refused my request. This could have been for a range of issues, but whatever the reason, it clearly affected the whole family.

My father had been Sunday School Superintendent at Higham for many years and both my mother and father were members of the choir. They had worked hard for the church in many ways. We decided as a family to resign and transfer to another church. It was a traumatic event. My father was in tears as he wrote out his resignation. My mother was cross and fumed, bearing in mind all the work that her father had done for the old Higham church and his work towards the scheme for the new church. The whole church council came to meet us and filled our front room trying to persuade us to change our minds. I played one or two pieces to show them that I was capable of playing the organ. But the meeting didn't result in any change of heart by either party.

I was saddened that my idea for gaining expertise as an organist was aborted. In retrospect, the new opportunities that came from the change of church were possibly for the best, as far as I was concerned. But I always felt guilty that, for my father, I had forced a change in his life that he might have preferred not to face.

## *Moving to a new church*

In 1953 we decided to move to Old Town Methodist Church on the outskirts of Barnsley, about two miles from our house in Higham. It was agreed that me and Dad would go to our first service one Sunday evening and report back to Mother. We went early, as we were not sure of the exact location of the church. We were the first of the congregation to enter the church, so we sat in the back row. A lady entered the church a few minutes after us. She turned around from the row in front of us and gave us a big smile. 'Good evening, and a warm welcome to you.' She was Mrs Denton and she could not have been aware how much the two of us needed that smile and that welcome. One smile and one sentence can often work miracles.

Old Town Church became a haven for my parents, and my father had his funeral service there. It was a very active church community, with many Christian skills and expertise available from the church members to support the running of the church.

There was a significant group of young people who were very talented, and they roped me into their activities. With a range of ages from early teens to Sixth Form students, and some who were in their first jobs, it was a unique group which contained good singers and excellent actors. One member was a music student who subsequently went to Cambridge University to take a music degree. He was known as an able arranger and composer of music. What he needed was someone to write the book and lyrics so that we could put on original shows. That is where I came in. We combined to produce a show called Amorous Flights, a nonsensical escapade in words and music exploiting the acting skills of the team playing unlikely characters in situations that would guarantee laughter from the church members. In this production I learned that the biggest laughs can come from unlikely words if the timing is right. A flamboyant character had the impressive sounding name of Sir Edward Fox-Waring. To engage with the rest of the cast he had the line, 'Just call me Ted'. In a subsequent discussion, another character, in a reply, said, 'It sounds very interesting, Ted'. It stopped the show each night, but the young man who spoke that line had the look and timing of a genuine actor.

During the next two years a group of us from the church went on two holidays. The first was by rail to Switzerland, for two weeks at Vitznau on Lake Lucerne. The highlight of the holiday was supposed to be a concert at a Music Festival in Lucerne. For some reason the seats had not been booked for us, even though we had paid for them. Discussions led to us being met by the Director of the Music Festival who led us to the best seats in the hall. We felt a bit conspicuous, as young people sitting among the great and the good of Lucerne, but we marvelled at the orchestra and its conductor, who I think was Rafael Kubilic.

The second holiday was on a boat on the Norfolk Broads. Some of the group had been on the Broads before, so it was assumed that they were familiar with handling the boats. In fact, none of them had any experience of boating, so I was put in charge. This worked alright until we were seeking to moor in Yarmouth when the engine cut out and wouldn't restart. We were drifting down the river, out of control. Fortunately, in our group was a startling blonde young lady, with a curvaceous figure. I instructed her to go up on deck and attract the attention of the harbour master. In her tight, white jumper, she had no difficulty in portraying a desirable lady in distress, and he came on board and sorted us out. This was a prime example of using any assets available to sort out your problems — a useful policy in life.

## *The facts of life*

Mining was a physical occupation in the 1950s. It is not surprising that the men talked about other physical aspects of their life. As a young man who had only had one lecture at the Grammar School on 'the facts of life' — which I did not understand — the conversations on sex were revealing and educational. There would be talk of the exploits in one of the seedier public houses in Barnsley and what the most notorious young women got up to. Young married men would describe the sex with their wife the previous night and postulate what they would do with her after the visit to the Club that night. Another young fellow gave his illustration: 'Me and my lass like to have quickies in the kitchen. I sneak up to her and lift her dress up and pull her knickers down. She leans forward over the sink and I go into her from behind. She knows how to get into position so that she gets all of him up to the hilt. She loves it that way. Of course, we need to make sure the kids are out. One day her mother knocked on the back door and came in just as we were hitting the high notes. I escaped upstairs. When she got into the kitchen she looked at the wife and said, "Oh you do look flushed and flustered, what have you been doing?" The wife was up for it and replied, "I have been cutting up the onions for a stew." The old bag will not get another chance. I will always make sure both doors are locked before we start.'

These tales, and many more, were not spoken with disrespect for their wives; it was more to emphasise that there was another physical side to their lives as well as working in the mine. There were still many fatalities in the coal mines each year, and it could have been a wish to savour what was available to them while they could have it, because the future carried its risks. After all, at every mine there were men in 'light jobs' — men crippled from accidents, whose chance to enjoy any sex life was zero.

However miners were a community of men with many talents and skills: there were gardeners whose experience could match any expert; and there were men who could sing. I once had a shift with two officials with me making a list of the equipment available for salvage in a part of the mine. While I was writing up the information, they were discussing the oratorios of the famous composers and jointly singing parts of the choruses — a remarkable underground experience. There were athletes: a shot putter, who worked in the blacksmith's shop at North Gawber Colliery and exercised his muscles by lifting one of his mates over his head each day; cricketers and footballers who were semi-professionals playing in the better leagues. There were brass band players, some of whom became classed among the foremost brass players in the country. There were devout men who were the backbone of local churches and others who served the community through different charities. There were other men who specialised in first aid work. They were an essential support to their local colliery, giving an immediate response to accidents, but they were also recognised for their expertise with the St John Ambulance Association at public events and competitions.

The most important quality of miners was their commitment to their colleagues. They were team players. If there were problems they would not walk away, they would give their all to work towards a solution.

### *Route to a wife*

At Easter 1956, I was commissioned to be the chauffeur at a wedding in Cawthorne. My mother had worked with the bridegroom's mother and our two families had maintained their friendship after their marriages and through the Second World War. I remember blackberry picking in the Cawthorne woods when we were all children.

The chauffeur's job had its challenges as the car was a Vauxhall Velox with steering column gear change. The trouble was that finding the middle gear of the three forward gears was a work of art. However, I got the bride to the church on time and completed all the other driving tasks. The chief bridesmaid at the wedding was the very smart, seventeen-year-old, younger sister of the groom. That day I began my quest to woo her. I offered driving lessons and devised other activities to establish contacts but there was strong competition. In my early days of courtship, I sensed that, as I was entering the back door of their bungalow, another beau was leaving by the front door. Even sixty years later, one man in Cawthorne still recalls getting into trouble with the teacher in the junior school for passing a note to his favourite girl in the class. Enid Herbert was always going to be a prize wife with many skills and attributes. It took three years of courtship before we were married at Easter, 1959.

The weekend before I started my final university exams, in 1984, before we were married, we had a short break at Whitby, supervised by her Grandma, who kept her beady eye on us. I look back on this break with amazement at the risk of taking two days off from swotting just before the start of the exams. But possibly it reflects Enid as the inspiration that she is, which helped me get a degree.

# Chapter 9

*University route; a terrifying interview*
*1954-1958*

The aim before going to university was to get a Coal Board Scholarship, as this would secure adequate finances to allow a university experience which was varied and free of financial debt. It was also possible to work back at the pit in the summer holidays, providing additional financial security.

All that was needed was to get through the Scholarship interview, which ranks as the most terrifying interview experience of my entire life. The interview panel consisted of the great and good of the NCB staff department, accompanied by the professors of all the university mining schools — in total in the teens of people — arranged at one long table across the room. Sat opposite them was a single chair in the middle of the room facing the interview panel. What damage could be expected from such an array of learned gentlemen?

Early in the interview I was questioned about the statement in my application form that I enjoyed reading history as a result of my A-level studies at Grammar School. 'What are you reading at the present time?'

I replied that I had just finished reading 'The Struggle for Europe' by Chester Wilmot, and was now reading 'Rommel's Papers' by Liddell Hart. This was a detailed analysis of the war in the North African desert. Immediately the questioning was taken up by Professor Whetton from Leeds University who, unknown to me at the time, had been a colonel under

Montgomery throughout the desert campaign. The rest of the interview was a discussion between the two of us on various aspects of desert warfare. When the questions advanced to issues that were unfamiliar to me, my answer was that I had not yet got to those matters in the book. Subsequently I learned that I was successful in getting a Coal Board Scholarship and the experience at the interview reflected the advantage of giving information on an application form that is unlikely to be presented by anyone else.

### A university place — just?

My plan was to go to Sheffield University and join a friend who was established there. However, I got an interview at Leeds University first, with Professor Whetton, whom I was already aware of from the Scholarship interview. The interview went well and he decided to offer me a place, but thought his list was full. He rang his secretary in the outer office and asked her if it was possible to fit me in. He got a positive answer, he said, as long as I decided on the spot. I was too immature to recognise his game, so I accepted the place to do a four-year honours degree course. I never regret that decision.

For my first year, I lived in digs and, after dropping off my luggage at the house, Dad left me to join the queue at the Freshers Conference. That was the loneliest moment of my life. I was terrified. This was in a new world and I did not fancy it. Fortunately, I made friends, particularly through the University Methodist Society, and quickly realised that the experience would be an interlude in my life that would affect everything that the future might throw at me.

The mining course in the first year was basic, but it became clear that my three years' experience in the industry would give me advantages over some of the students who were entering the course straight from school. I was able to play football, usually three games a week, initially as a centre forward but in later years as the goalkeeper. I joined other societies, but I was not a major resident in the bar, nor did I join a group of mining students who spent their Friday evenings touring the pubs of Leeds, sampling their wares.

My landlord decided he needed payment through the Easter holidays to guarantee availability of my room for the summer term. I declined his offer and one of my friends in the mining department found me a place in his digs, where the owner had constructed an attic room. His wife had been a boardinghouse keeper in Blackpool and she fed us so well that when I went back to the pit during the summer holidays, they did not recognise me as I had put on so much weight.

The second year of the degree course presented two academic challenges. Firstly, a course in the Maths Department introduced us to levels of mathematical thinking way beyond the A-level maths course. The change was such that one or two students from the Mining Department withdrew from the lectures, and opted to take an Ordinary Degree, rather than an Honours Degree. I persevered and, through brilliant teaching by the professor concerned, managed to pass the exams.

That year we also started studies in geology, and these lectures were taken in the Geology Department. It seemed that the Geology Department staff had a jaundiced view of mining engineers and took delight in failing any mining student who was in any way suspect in the exams. In retrospect this seems a little unfair. The one geological fact that sticks in my mind was a statement made by Professor Versey, the head of department, in an introductory lecture he gave. He said, 'There are many uncertainties in geology, but of one thing I am quite certain: there is no oil in the North Sea.' I was relieved to get through the geology exams. Much later in my career I realised that the oil and gas fields are below the coal measures in the North Sea, at significant depth.

*Sadler Hall*

For the last three years at university my accommodation was in Sadler Hall, in Adel. It was a small men's hall of residence, with 35 members in an old stone-build house set in beautiful gardens. The warden was Dr Higginson, who was an educationalist, supported by his wife. He was a tall, academic looking man, to whom life was a serious business. His wife was a lisle stocking-lady whose physique suggested sporting prowess, but her major role was as the car driver of the marriage. She was prone to ambiguous comments at the formal dinners that we had each week, when special guests were invited. On one occasion, describing holidays on farms in the country, she said, 'There is nothing so wonderful as to be woken up first thing in the morning by the farmer's cock!'

I remember the outrage among fellow students at the Suez invasion in 1956. During the same year there was the suppression of the Hungary uprising by the Russian army. Leeds was a host city for many of the students and young people who were involved in the uprising. There were some amazing stories of the Russian tactics reported by some of the students who had managed to escape. Probably the most touching one was a young married couple who became separated during the escape process but were reunited on a platform at Leeds station. What a joy for them and the people who had organised the rescue. We decided to finance one Hungarian student through his university course and he joined us in Sadler Hall. Of the many fund-raising initiatives, the most memorable one was carol singing at the homes of the wealthy residents of Adel. Much money was collected, but also much alcohol was consumed at each house, such that the tour had to be concluded early, as the level of inebriation among the members precluded the ability to sing carols.

In my final year I was elected President of Sadler Hall. This was a great honour and gave access to the wider aspects of university life. At that time, 1957/58, there were about 2,500 students at the university (compared with over 25,000 in 2011). The Presidents of the halls met the Vice-Chancellor and other senior management of the university. The President also had responsibilities for the members of hall and their behaviour. It was a changed circumstance in that year as many students were entering university straight from school without doing National Service first. The average age seemed to drop significantly. One new member, who appeared late, via America, was an African, Vincent Nuizugbo, who presented me with some problems. One night he came into the common room and announced, 'I have got music, food and alcohol but I just can't seduce this young lady.' There were some startled looks from the younger members of hall. Of more serious concern was his financial problems that arose a few weeks later. The money to finance him had not been forthcoming to the local bank. The Warden asked me to warn our students not to lend him any money. With no end to the problem in sight, the Warden called Nuizugbo into his office to discuss his financial situation. After the Warden had opened the discussions Nuizugbo gave his famous reply, 'But, Warden, we are gentlemen, we do not discuss such problems.'

Dr Higginson later told me of that meeting and he added, prophetically, 'That just shows why the Africans are not ready to rule themselves at this stage.'

## A first-class honours degree?

The final year exams concluded with two special papers. As I entered the department for the final flourish I was met by Dr John King, who was senior lecturer. He greeted me with a statement that immediately super-charged my nerves. 'You are doing very well. I hope that you can keep going at that level.' I realised that I might have a chance of getting a first-class honours degree if I did well in the special papers. When the results came out, I had indeed passed with first-class honours but what was that worth? I decided to go home to tell the good news to my parents. As my mother was visiting a friend of hers in Wakefield, I called there to see her. I parked my Lambretta scooter and entered the house carrying my helmet.

My mother was in the full flow of a conversation with her friend, which she was reluctant to close. At one pause to draw breath she looked across at me and said, 'Did you get a first?'

I replied, 'Yes,' and she resumed her conversation without any comment.

And after all, the piece of paper recording the degree is the smallest certificate I have ever received.

## *Directed Practical Trainee, 1958/59*

The Coal Board ran a scheme for initiating graduates into the industry. It was normally a three-year scheme but, because of my pre-university experience, mine was reduced to one year. It was designed to show how the industry functioned, with short, snap-shot periods in the various specialist departments and services of the organisation. I spent a few weeks with a mechanical engineer at a pit, also a short time was allocated in the group wages office setting up the wages to pay out to nearly three thousand men. I had never seen so much money in my life; the security of the building was similar to that of a bank. There were periods on other departments, but the most instructive time was that given to shadowing a colliery manger and a colliery under-manager. There I got a clear view of the responsibility and the range of issues that they had to resolve.

As I had the required practical experience I was able to take the Mine Managers Certificate examination while I was on the Directed Practical Trainee course. This was a notoriously competitive exam. You looked at the two students beside you and you knew that only one of you could expect to gain a pass. The problem was the Legislation Examination. The Coal Mines Act and Regulations were extensive and detailed. They were specific on who had responsibilities and what action should be taken in any particular situation. The questions were often formed around incidents that were described. At my first attempt I wrote a commentary with my answers, but I found that what they wanted was quotes from the Act and Regulations. I amended my style and got through the second time. Some people took the exam into the teens of times.

Also, while on the Directed Practical Trainee scheme I passed some exams administered by the Institution of Mining Engineers, which qualified me to become a Chartered Engineer when I had gained the required management experience.

## A big life event and a residential course

On 30 March 1959, Enid and I were married in Cawthorne Church, surrounded by our friends and family. We were fortunate to have a house to move into. I'd said to Enid, on our engagement, how convenient it would be if my uncle, who worked in Higham Cooperative shop, was to get the manager's job. He could then move into the house which was part of the shop, and we could get married and move into his house. That is exactly what happened a few months later. So, we never had to go house-hunting!

*Enid and I on our wedding day, 30 March 1959*

Near the end of my Direct Practice Trainee scheme I joined a residential course at Manchester University. This was immediately after I returned from my honeymoon. The journey over the moors to Manchester was in torrential rain and I was like the proverbial drowned rat when I arrived on my Lambretta. The course was interesting and showed us the different mining conditions (thick, deep, highly sloping seams) compared to ours in Yorkshire. It included much essential information about the coal industry and was attended and addressed by a bevy of senior managers from Lancashire. I was given the task of thanking the Course Director and his team for the excellent way they had run the course. I began my short speech by stating that, having come straight from my honeymoon, I was the member who least wanted to come at the beginning of the course and the keenest to get away at the end! I said that my remarks would be short, but would be very sincere.

# Chapter 10

*First step on the ladder*

At the end of the Directed Practical Trainee scheme I was qualified and ready to go as an underground official in the coal industry. However it was with some trepidation that I was taken to meet the new colliery manager at North Gawber Colliery, Mr C W Turner. He turned around and gave me a big smile and said he had heard of me (he must have been reading the book 'How to Make Friends and Influence People'!) It was to be the start of my long association with the industry and with the Institution of Mining Engineers. What was more important, he appointed me as a Deputy Grade 1, in view of my qualifications, rather than the normal starting grade of Deputy Grade 2. The pay was £30 per week for the Grade 1 Deputy instead of £29 per week for the Grade 2 Deputy. In my newly married state, without a stair carpet and other household furnishing and appliances, the extra cash, added to Enid's salary as a junior school teacher, helped to establish our home.

Charlie Turner was a progressive and successful colliery manager. He was a mentor to me and he passed on tips from his style of management. I always recall that he said you should not play the ace in any negotiations if you could win with the nine. He also took me to the public enquiry on a mining incident which showed the legal powers of the many interested parties in finding the truth and apportioning blame. It is a lonely role for a colliery manager to be in the witness box. However, Charlie initially put me to work with an old stage under-manager who regarded me as just another official and under him I might have made no progress at all in my career. However, I was soon put in charge of several teams of men to develop a new coal face and we made good progress, somewhat embarrassing the under-manager by the number of tubs of dirt we were producing each day.

## *Shaft sinking*

My next role was with a shaft sinking project at Woolley Colliery. Woolley was an extensive mine and the production was being affected by the limiting capacity of the mine fan. The scheme was to provide a new large shaft on a site some distance from the Woolley colliery mine site. A large fan was installed, and it was used exclusively as a ventilation shaft. The scheme was a complete success when it became operational.

*View from the shaft bottom of the cactus grab used for filling out the material excavated in the sinking process.*

The key means of travelling the shaft by men and materials, and bringing the excavated debris to the surface, were the hoppitts. These were cylindrical containers, four feet in diameter and six feet deep. At the top rim of each hoppitt there were three chains which were coupled to a circular link, which was used to connect to the winding rope. When the hoppitt was standing free in the shaft bottom, the chains were draped down its side so that the debris could be loaded into it. When it was full, the winding rope was lowered, and the chains were connected to it through the circular link. An important action then was for the hoppitt to be stabilised in the correct vertical line to travel through the four-deck scaffold and up to the surface.

Under the bottom deck of the scaffold the machine for loading the dirt was mounted. It was on an arm which was secured to the outer rim of the scaffold and it could be rotated around the shaft and moved backward and forward along its arm. The operator was in a central position of the machine and rotated with it so that he could see where it was working. The loading machine took the form of a compressed air-operated cactus grab which picked up the dirt and carried it over and dropped it into the hoppitts. The skills of the operators varied immensely, and the best ones effectively threw the dirt from the grab into the hoppitt. The poorer operators often took some time manoeuvring the grab to align it over the hoppitt. The two hoppitts, one on either side of the shaft, were raised to the surface by independent winding engines. They were unloaded onto a chute by the hoppitt being lowered by the winding engine while its base was held in a fixed position. During this tipping process the doors over the shaft top were closed to avoid any materials falling down the shaft.

To drill the holes at the base of the shaft a drilling basket was lowered from the surface with up to ten drills on it which had been prepared and serviced in the engineering shop. Over 100 holes were drilled to a depth of eight feet and to a pattern supervised by the master sinker to ensure that the finished diameter of the shaft was correct after the concrete walls were installed.

As Deputy I had direct control of the shot-firing operation on my shift. This was extensive as it involved charging over 100 boreholes at the base of the shaft sinking, each hole being eight feet long, and it was essential to get the explosives to the base of the hole. The explosive was polar ammon gelignite, a strong explosive especially prepared for shaft sinking projects. The delay detonators ensured that the centre of the shaft fired first and, with millisecond delays, the outer rings of holes were then fired. When the holes were charged and connected to the firing cable the scaffold was raised so that it was well clear of any flying debris. Everyone was brought out of the shaft and the klaxon sounded before the round was fired. It was a mighty explosion and a real thrill to turn the key as the Deputy.

It was normal in secure ground to sink about 60 feet and then stop for concreting. To do this a curb ring was lowered and set in place by the surveyor to ensure absolute verticality of the shaft. Concreting then took place in lifts until it connected to the shaft wall already completed. The concrete was poured down a steel pipe to a dash pot near the bottom to take the pressure, and then it was poured with flexible pipes behind the steel formations installed to hold the concrete. The concrete was vibrated to ensure it was consistent.

Riding the hoppitt through the shaft and landing onto the platforms or into the shaft bottom called for some nerve, with no fear of heights. I had no problem with this.

The set up at Woolley for shaft sinking remained as standard practice for many years. The ten shafts at Selby had similar four-deck scaffolds and the winding engines, shaft furnishing systems, mucking units and concreting arrangements were the same as used at Woolley.

Two incidents during this period are of interest. Firstly, there were no restrictions at that time on the operations, so we fired rounds of shots at any time during the twenty-four hours. When we were near the surface and passing through a band of sandstone rock I realised that there would be shock waves affecting a farmhouse quite near the sinking. I often wondered, when I fired in the middle of the night, what this did to the sleeping habits of the farmer and his wife; I hoped that it threw them nearer together! Secondly as a young man I had been told that one must never refuse a drink from a Scotsman. Cementation brought in a new master-sinker called

Walter Connelly from Scotland. After a trip home one weekend he returned restocked with whisky. He requested me to join him. 'Just come over to the cabin, Trevor, my boy, and we'll have a wee dram together.' This was it; I dare not refuse. In his cabin I was shocked to see him pour a great slug of whisky into a half-pint pot. I took a deep breath and tried to smile. We touched mugs and I drank it down. Going down the shaft some time later to carry out a routine inspection, I clung onto the hoppitt chains as the shaft seemed to be going around in circles. It was the last time I ever drank whisky — even with a Scotsman!

## An unbecoming father, 1960

It was after a nightshift on the shaft sinking that I came home to find Enid on the bathroom floor, clearly in the process of giving birth to our first child. The birth was not expected for a few weeks yet. I got her wrapped up with the things she had been getting together for the expected birth and drove into Barnsley to the maternity home. At red traffic lights in the middle of the town, Enid said, 'Don't stop now!' I then had to leave her to face the challenge of childbirth, purely in the hands of strangers. It seemed a cruel imposition to place on a wife.

The baby, a boy, was born within two hours of Enid entering the nursing home, I found out when I rang up later in the morning to see how she was getting on. I then went and told her mother the good news and she suggested that we should go into town and buy her some flowers. By the time all this had taken place it was early afternoon and I was exhausted. I went to bed and set the alarm clock to wake me in time to go to the evening visiting. When I woke up, I looked at the clock and was aghast to see that I had slept through the alarm and the visiting time was over. I drove into town and begged — really begged — with the nurse in charge that I should be allowed to spend a few minutes with my wife. Fortunately she agreed, but I carry the sense of guilt that I had let my wife down on a day when she needed support. Visions of her lying in bed, all alone, while the other new mums had their husbands at their bedsides, still send shivers of horror down my spine.

We named our son Brynnen David Massey. We have no Welsh connections, but Enid taught a boy called Brynnen in her junior school class and I had liked the interesting stories that he wrote in some of his compositions. My mother was not thrilled by this name. She suggested that it would be shortened to Bryn. I replied that I could think of other names that are shortened to much less complimentary nicknames.

### *Playing village cricket*

As I have already mentioned, our first house when we were married was adjacent to my original home. The house had belonged to my uncle, who had been promoted to manager of the Higham Cooperative store, which had a smart house connected to the shop. It was very convenient, being where we were, as the children could climb over the hedge at the bottom of our back garden and they were in the smallholding of their grandparents. My mother always claimed that she was able to pass on her baking and housekeeping skills to the children, which she believed to be essential to their education. This would have not been as easy had we lived any distance away.

The house was also very convenient for access to the village cricket field, which was only two hundred yards away. It must have one of the best outlooks of any cricket pitch in the country; looking west there is a view of fifteen miles through rolling countryside up into the Pennine hills, with the Emley Moor television mast sticking up like a great finger into the sky. The field was sculptured into the hillside, the outfield dipping away gently from the pavilion to the playing square and then the rest of the ground was reasonably level, with wooden railway sleepers as seats around the boundary. The playing area was well maintained by the groundsman to give wickets that played true.

The club fielded two teams in local leagues. I played for about ten years, starting in the late 1950s until we moved to a new house at Woolley Colliery in the late 1960s. My role for most of that time was as opening batsman.

Occasionally I achieved a score of over fifty, and it was then traditional to have a collection around the field, so I went home with a pocketful of small coins. For two years I was captain of the first team — a role which introduced me to the politics of the game. The Sunday meetings of the committee in The Engineer public house exposed me to traditions of team selection and evaluations of individual performances which often didn't coincide with mine, as captain.

There were some interesting umpiring incidents. One umpire came from Dewsbury and he travelled by bus. He was a small man and was generally fair in his decision making. However, there was only one bus each hour to Dewsbury, which he caught in the next village and that was about a ten-minute walk from the cricket field. The team batting second had to realise that if a number of wickets had fallen, he would consult his watch and calculate when he must leave to catch the next bus. From then on, any appeal for a leg-before-wicket or a run-out would get a positive answer no matter how dubious the request. If that umpire was on duty it paid to bat first if you had the choice. On another occasion in a needle match between Higham and Silkstone, a Silkstone batsman was given out for a catch behind the wicket, but he was adamant that he had not touched the ball. He refused to leave the wicket and, with his partner, sat down in the middle of the pitch. This presented a classic strike situation: there was no chance of a compromise as the umpire would not change his decision; the game might have to be abandoned. But Higham were in a good position to win the game. So, the Higham captain, not me, agreed that the batsman could continue in order that Higham could win the match, which they duly did. This is an interesting example of the sort of cases that can arise in industrial relations.

In one season I hit a bad patch when I was getting out for a few runs in each innings before I had settled. I talked about this with a fellow undermanager at Woolley, who also played in the Higham team, as we travelled to an away match in a village near Wakefield. There was no solution to the problem other than to keep trying and watch the ball. The local team had a young man as their opening fast bowler who was known to be very good, so I went to the wicket expecting a short innings. While the young man was very fast, his direction was variable, and I managed to hook him to the leg boundary for several fours and the odd six. Suddenly, my confidence grew, and I ended up with a score of 114-not-out. Despite all the exercise I got at that time with frequent underground pit visits, at the end of the match my body really ached, and I needed a hot bath when I got home. That score helped me to win the batting averages for that season.

Once, there was a meeting with one of my cricket colleagues in strange circumstances. In 1956, my father decided to declare some additional income from the market garden he was running as well as his main job. The Inland Revenue decided to have a full investigation, thinking that they were onto something big. They asked numerous questions and built up a profile of the family's income and expenditure to demonstrate that the income must have been much higher than he had declared. He was able to refute many of their suggestions and they didn't realise that he was so busy working that he did not have time to spend his discretionary surpluses. Their estimate of holiday spending was much too high and fortunately my Dad had receipts for some holidays which were very low as they always hired a room and provided their own food. It was known as 'keeping yourself', which has long since faded from the options offered by seaside landladies. After much investigation, my Dad was called in to meet the head of the Inland Revenue office in Barnsley.

I came home from university to join my parents for the meeting and it was a surprise to see my cricketing colleague supporting his boss on the case. The meeting did not present any figures that were acceptable to our family. It was noteworthy for an outburst from my mother who banged her umbrella on the desk to emphasise that they always paid their taxes and worked hard for their money. The manager suggested that a payment of just over £100 should be paid to settle the matter, and this was agreed. Subsequently an accountant was appointed to prepare annual accounts for the market garden and my father was then able to claim numerous allowances so that he got a rebate from the tax he had paid with his full-time job. If our case was an example of revenue investigations, then I think my colleague should have concentrated on his all-rounder role on the cricket field.

## Colliery Overman, Wharncliffe Woodmoor 1 2 3, 1960-1961

After my spell on the shaft sinking there were several changes in the management of the No.6 Area of the Yorkshire coalfield. Under H J, Atkinson the No.6 Area in Yorkshire had been very successful financially. It was normally second in financial terms to one Area in Nottingham. The results were achieved utilising good workmen who were still hand-filling coal. There was limited investment in mechanised coal getting. A new Area General Manager, C V Peake, was appointed with a mandate to introduce mechanised coal production. Charlie Turner, my mentor, was also promoted to Group Manager over several pits in the area. He moved me to Wharncliffe Woodmoor 1 2 3 pit as a colliery overman, which was the next step up from colliery deputy. The pit was introducing mechanised coal faces and had installed conveyors to collect all the coal to one loading point in the pit bottom. There the coal was filled into pit tubs and wound to the surface.

It was at this pit that I learned the essential understanding of manpower deployment underground. All the men on the dayshift collected along a roadway in the pit bottom and the overman had to direct them to each of the coal faces and the other development work throughout the pit. When everyone had moved off, he then went to a small underground office and filled in a book listing where everyone was deployed underground. This was totalled up and the figure was compared with the number of men recorded by the banksman as being underground for that shift. It was a good system laid down by the manager of the pit at that time.

The experience with mechanised coal faces was mixed. As the floor strata under the coal seam was soft and the supports sank into it, it was decided to take up some of the soft floor so that the supports were set on a firmer base. The roof conditions improved, but the amount of dirt sent out of the pit outstripped the coal preparation plant and affected the saleable output.

The central conveying scheme also produced problems. There were extensive delays on the coal faces as the shaft could not clear the coal fast enough. The size of these delays was demonstrated when, on the day before a holiday, there was a poor attendance and some coalfaces didn't operate at all. The mechanised coalface was manned up and produced four shears in the shift. It indicated the potential of the mechanised faces, as it was twice the normal shift performance.

The deputy and myself were called into the manager's office to be congratulated on the shift's result.

# Chapter 11

*First steps in senior management*
*1961-1965*

1961 saw my first move into senior management when I was appointed undermanager at Wharncliffe Woodmoor 4 & 5 pit. I was normally functioning on the afternoon shift to support Jack Sykes who was undermanager on the day shift. Jack had bad asthma and whenever he was off work I moved to take his place on the dayshift.

The pit worked three seams and all the coal was transported to the surface on a central conveying scheme up a surface drift. The mine was not noted for good industrial relations.

My time there was memorable for my first experience of being a colliery manager. Both Jack Sykes and the colliery manager, Frank Ramsden, were on holiday when the pit took its main holiday week. I was given a temporary colliery manager appointment by Charlie Turner, and the written document spelled out in detail my responsibilities.

Two incidents that week needed my direct involvement. A contractor was on site to excavate a connection from the surface to the shaft side to form a new fan drift. As the excavator made progress it was stopped by a solid brick structure that had been a part of some foundations for an earlier coking plant. The operator approached me and asked if I could arrange to 'soften' the brick structure as his machine could not make any further progress. I called the overman, who was in charge of the underground operations, to investigate the problem. He said he would arrange to soften the brick obstruction with some explosives. I sat in the manager's office considering that this might be the shortest tenure of a colliery manager in the history of coal mining. I had my head in my hands when there was a violent explosion and I heard a brick from the structure land in the pit yard near the office. True to his word the overman had indeed 'softened' the obstruction. There were no complications, although the people in the row of houses near the pit may have thought that the explosion indicated the start of a major military action!

The second incident occurred on the Friday at the end of the holiday week. The foreman fitter came to tell me that, along with an apprentice, he was going to collect all the pit ponies that had been on their holiday in a nearby field. At lunchtime, two bedraggled men returned to the pit yard, their boiler suits tied around their waists and sweat dropping from their chins. The pit ponies wished to extend their holiday! They had not caught a single one of them. It was soon realised that it would be necessary to build a corral in the field and use it to catch the ponies.

There were six coal faces at the mine, all hand-filled, and the coaling shift was days one week and afternoons the following week. Human relations were not easy, and on some days there would be a pit bottom with over 100 fillers refusing to go to work until some issue was resolved. It is a challenging prospect to walk out into such a situation with most of the men shuffling about and not prepared to look you in the eye. You have to seek out the leaders and unravel the real issue before you can concoct some possible steps towards a resolution of the problem, in order that the men would go to work.

At this time, we had no telephone in our house, so I had to go up the village each night to a telephone box to speak to the officials for the results of their shift and the plans for the next shift. Getting a telephone into the house was complicated, but without it was like walking blindfold on a dark night.

It was around this time that I bought a new car — a Vauxhall Victor — which became important for transporting the family, particularly on holidays. Our family became complete with a brother for Brynnen, Adrian, born in March 1962 and a sister, Dawn, born in October 1963. The new car had a roof rack, which was useful as we went camping at that time. As my working day started very early, at 5.30am each morning, our holiday practice was to pack the car the night before and, in the morning, load the three children in their night clothes into the back of the car, where they could continue their sleep, and start the journey. Our normal destination was Northumberland, so we would break the journey well up the A1 and dress the children there. Enid and I regarded this as a sensible arrangement until we heard the children discussing this practice many years later when they were home from university. They described it as a cruel arrangement which should have been referred to the NSPCC!

## *Back to where it all started, Woolley Colliery, 1963*

In 1963 I returned to Woolley Colliery as undermanager of the Parkgate seam, the very part of the pit I had visited in 1951. The men and the officials at Woolley were very good and the trade union leaders were keen for the pit to be successful. My uncle worked on the coal face in the Parkgate seam, but few people realised he was related to me.

One person who worked for me, who became very famous, was Arthur Scargill. While I was climbing the management tree, Arthur was climbing the National Union of Mineworkers (NUM) tree. He didn't cause me much trouble. He worked as part of a small salvage team which included his father —a communist — along with another known communist called Loll Parry, and the son of Alwyn Machin who was President of the Yorkshire NUM. They presented a lethal political team.

The way the issue was addressed was to put the team on a salvage job in a distant part of the mine where there were no coal faces in operation. They were also deployed on a split shift, which separated them from the main shift changes when the majority of the men were around. It worked alright, but the icing on the cake was to put a deputy called Arnold Hanks in charge of them. Arnold was a very good deputy, but he was as deaf as a post and he had to wear a hearing aid. If Arthur and Co. started preaching their philosophy, Arnold just turned off his hearing aid and he was immune to any of their arguments.

A couple of years later, when Arthur was seeking an official role in Woolley NUM branch, he applied to have day release to study industrial relations management at The Northern College. His request was granted, which was an advantage to the colliery management as Arthur was not involved for a period in the union issues at the mine. It might, much later, prove to be a more serious problem for the country, as it was rumoured that on that course new militant techniques for organising strikes and picketing were on the agenda.

Each colliery had its own union branch and the key figures on the branch committee were the union President, the Secretary, the Treasurer and the Delegate. The President was usually a powerful personality who could handle the union meetings and could hold his own, as far as the men were concerned, in meetings with the management. The Secretary did all the work and handled a lot of the men's employment and pay issues. The Treasurer looked after the branch funds, which could be quite extensive and include property and other assets. The Delegate was the person who represented the branch at the meetings of the Yorkshire NUM, held in their offices on Huddersfield Road in Barnsley.

In the 1950s the Yorkshire NUM was a moderate part of the National Union of Mineworkers, on a par with the Nottingham miners' union. The communist party realised that if they were to get the NUM as a whole to be a militant union then they had to change one of the larger coalfield unions from being moderate to being militant. They decided to go for the Yorkshire coalfield. From a club on the outskirts of Cudworth near Barnsley, a communist from Scotland created a cell to effect the change. The way it was achieved was to place a person with militant leanings into the delegate position at most of the pits in Yorkshire. That was how Arthur Scargill managed to get in the Woolley NUM branch.

Arthur was not the dominant character in the Woolley NUM; that role was played by Elijah Benn, the Treasurer. He wanted his hands on the money and he certainly achieved that. As well as being treasurer for the NUM branch, he was treasurer for the home coal scheme, he was treasurer for the canteen committee and he was treasurer for the Woolley Miners Welfare Club. He always walked about with the largest roll of bank notes that you could ever imagine in his pocket.

As far as Elijah was concerned, money meant power and he used that power. He was known to deliver coal free to widows who were having a hard time. He used it to good effect during the elections for the NUM branch at Woolley. The election took place after the retired miners tea and concert at the Miners Welfare. Retired miners had a vote, so buying free drinks for them was an opportunity to discuss voting possibilities with them. He also used his bank role to outflank me on one or two occasions. We would be arguing some wage dispute with a workman and I would explain that it was impossible for me to make any payments under the agreements. Elijah's next move was to ask for an ex-gratia payment to settle the matter. If I refused this, on one or two occasions he played his ace: he would take the roll of bank notes out of his pocket and peal a few off and give them to the workman. 'Here lad, take these. I'll pay thee, and I'll sort it out with the Boss later.' There was no answer to that move!

But Elijah had a heart as well. Sometime later, when I was deputy manager of the mine, I remember meeting with the union team to sort out that week's wage issues on the Friday of the Abervan disaster. We had picked up scraps of news during the day, but in the afternoon the scale of the disaster was becoming clear. We realised that the lagoon slurry had flooded down the hillside and the school and the children had been buried. Elijah folded up the pay notes and put them in his pocket. 'We are not discussing these notes today when them poor bairns are buried in that school.'

One case worth noting is when Arthur Scargill was representing John Thomson, a workman in the Parkgate seam, who had hit the colliery overman, Gordon Mason. This was potentially a dynamite situation as there were two unions involved — the NUM and the official's union, NACODS (National Association Colliery Overmen Deputies and Shotfirers). Arthur saw the undermanager, Mr Caswell, and discussed how they should handle the case to limit the damage. It was agreed that John would be brought out of the pit straightaway and Arthur would take him to see Mr Caswell. They would establish the facts. It would emerge that John had had an argument with Gordon Mason, they had raised their voices at one another, they had shouted at one another, they had pointed their fingers at one another in the heat of the argument and in doing this John had accidentally caught Gordon on the side of his face, for which he was sorry and was prepared to apologise.

Arthur accepted this plan and he met John as soon as he came out of the pit. He emphasised to him that it was a serious issue, but assured John that he thought it could be sorted out. He was taking him to see Mr Caswell and John had to agree with everything Mr Caswell said.

Imagine the scene! A big workman, in his pit clothes, standing in front of the undermanager in his office. The conversation went something like this:

'I understand that you have been working on 14's coal face in the Parkgate seam today, John.'

'Yes, Mr Caswell.'

'And I understand that you have had a fall-out with Gordon Mason?'

'Yes, Mr Caswell.'

'And I understand that during the argument there was some shouting and it got a bit noisy?'

'Yes, Mr Caswell.'

'And I understand, John, that as well as shouting, you were also pointing at one another.'

'Yes, Mr Caswell.'

'And I'm told, John, that when you were pointing at one another you caught Gordon Mason on the side of his face?'

'Yes, Mr Caswell, then I hit him with my left hand and thumped him with my right hand!'

A case lost for Arthur!

Woolley Colliery worked four seams out of the sequence of seams in the Yorkshire coalfield. The Parkgate seam was over six feet thick and had a seam of dirt splitting sections of the seam. It was therefore very productive and was a major constituent of the total pit output. Three other seams were worked at the pit: the Thorncliffe seam, which was medium section, but with limited reserves; the Lidgett seam and the Silkstone seam were both thin seams, around two feet eight inches thick.

The coal face technology of the Parkgate seam was an early version of mechanisation. The shearer was a 200 horsepower machine with a 76-inch diameter cutter drum. The powered roof supports were a sequence through the coalface of two leg supports and three leg supports. The hydraulic system to operate the supports used mineral oil, so whenever there was any burst in the hydraulic hoses the pump was switched off and the repair carried out otherwise all the oil would be lost. The method of working was for the shearer to cut along the length of the coal face. As the machine advanced, the two leg supports were pulled over to give immediate support to the exposed roof. When the shearer reached the end of the face a section of the cutting drum was removed, which allowed the shearer disc to pass under the two leg supports. The shearer then traversed back to the other end of the coalface using its plough to fill the loose coal onto the face conveyor. The face conveyor was rammed over and lined up as near the solid coal as possible and the three leg supports advanced to complete the cycle.

When cutting, the stresses on the shearer disc were very high and it was not unusual, from time to time, to have a few of the picks on the disc breaking off, such that it had to be brought out of the pit to be welded up. The teams on the coal face were very cooperative and on some days we could arrange the men to work on a later afternoon shift and stay until they had completed the normal number of shears for the day, despite the disc repairs.

When I was undermanager, there was a major development for a new coal face that would be 250 yards in length. There were three sets of men opening out the face, which involved nearly 600 yards of total drivage. The teams were making good progress but when the Senior District Inspector visited the developments on a speculative call, he found that we had omitted to apply the required stone dust to suppress the fire risk. I was reprimanded but the matter was quickly resolved. When he came back on a follow-up visit, everything, including the workmen, were covered in stone dust. We generally worked amicably with the mine inspectors and on their visits they were treated to a good lunch while their car was cleaned and a bag of firewood put in the boot for when they left.

Throughout the coal industry there were major reserves in thin seams and this was often high-quality coal. A push was clearly needed to mechanise these thin seams so that they could compete with the thicker coal seams.

In 1964 I was transferred to the Silkstone seam at Woolley, which had extensive reserves. The seam had a very good sandstone roof, giving ideal strata conditions for thin seam mechanisation. After examination by senior executives of the NCB it was decided to try a Remotely Operated Longwall Face (ROLF) on 3's coal face in the Silkstone seam.

The aim was to reduce the number of men working on the coal face where the extraction would be 30 inches. The powered supports would be operated from a panel in the roadway, either automatically as the coal cutter moved along the coal face, or remotely on the instruction of one workman travelling along the coalface. A new type of machine was developed for the face which was pulled along by a continuous chain. The machine was steered by a ram mechanism behind the cutting head, which applied pressure to lift or lower the cutting head to stay in the seam. It soon became clear that the machine could not be steered within the coal seam and, from time to time, it locked fast and could not be moved in either direction. This led to a tragic accident.

The machine locked fast in the middle of a night-shift and the fitter, Geoff Gledhill, decided to use the time to measure for a new machine guard. The access to get the measurement was very difficult, due to the limited extraction. Suddenly the machine released and moved several inches forward crushing his head and killing him outright. It was a very messy business. The young overman in charge of the coal face was devastated by the accident and really went into shock. Fortunately, the deputy for the district had been a medical orderly in the navy during the war and had experience of clearing up after battle casualties, so he sorted things out underground. I went back to the pit and was involved in the detailed work with the mines inspectorate, the police, the trade unions and the next of kin. The Colliery Manager decided to go underground with the mines inspector so, after the dayshift operations at the pit were resolved, I went home for some breakfast. As I was in the living room with Enid, my three kids came bounding down the stairs in their pyjamas for the start of another day and it hit me: my three kids — two boys and a girl — were the same ages as Geoff Gledhill's kids. There would be three more kids coming down the stairs to start another day, but they would have lost a great part of their life. Moments like that drastically impact on your experience and are never forgotten.

The machine for the coal face was replaced by a shearer and the extraction was increased to 32 inches. The coal face operated successfully with the standard thin seam shearer, but the ROLF concept was not widely applied in the industry.

### *Deputy Manager, Woolley and Grimethorpe, 1965-1969*

The Woolley management organisation, which used to have two managers — Woolley A and Woolley B — was changed to have a General Manager and a Deputy Manager. Calvin Round was promoted to General Manager and I was appointed his Deputy. The change affected our domestic arrangements as well as my role in the pit. A large house had been built in 1938 on the hillside above the pit by the engineer for the mine at that time, for his own use. It had a one-acre, fenced, private garden with greenhouses and a tennis court. Calvin did not want to live there, as he considered it too near the pit, so it was offered to me. It was very convenient as the school for the children was on the hillside 100 yards from the front gate. I could walk down, through the back gate, to my office in five minutes. The children could be in one bedroom where they could see the three operating shafts and they could easily let me know if the wheels were going round to bring the coal to the surface. We had a gardener who provided us with fresh fruit and vegetables, and he lit the fires each morning to warm the house, as there was no central heating. We had a colliery phone in the house, so I could ring anywhere on the surface or underground if there were problems. Occasionally, Enid got some stray calls asking for her go to some specific problem underground, so she developed an icy response stating that she had no intention of going underground, whatever the problem!

We developed a routine at the end of my day at the mine. I would ring Enid and ask her to send the three children and our Alsatian dog down to meet me and we walked together back up to the house discussing the children's accounts of their day along the way. The dog, Flix, appreciated this routine, but on some occasions she would escape out of the back gate and come down to the pit on her own. The manager's clerk would come into the office and say, 'The dog is here, Boss.' My reply would be 'Send her in.' She would sit at the side of my desk until I was ready to take her back home. Having an Alsatian dog sitting at the side of the desk could be considered an unfair advantage when negotiating with NUM officials!

In the 1960s there were still pit ponies widely used underground throughout Woolley colliery. They were well looked after and any accident to a pony was fully investigated by the mines inspectorate. A memorable sight was the evening before the annual pit holidays. All the village children would crowd into the pit yard to see the ponies being brought to the surface to go into an adjacent field for their holidays.

In the Lidgett seam we had mechanised coal faces using trepanners. (Their cutting mechanism was the same as that used by surgeons to cut through the skull for brain operations.) The machines worked hard at cutting the coal, but were not very reliable. When the large centre section of the machine failed, we had the problem of getting the new machine part to the coal face. There was only one horse in the Lidgett seam that was strong enough to carry out the task, and it was a struggle even for him. Whenever we had a breakdown I would go back underground with a bag of sweets to entice the horse from the stables. With two men pushing at the back of the loaded tram we would tackle the journey in stages. At each stop for a rest I would bribe the horse with sweets and encourage him to complete his task on the basis that he would then have a day off in the stables. He was a wonderful horse!

The partnership between Calvin Round as General Manager and myself as Deputy Manager worked very well. I decided that it was good experience for me to do things such as negotiating with the unions, which he delegated to me, while he often chased up trivial issues that I thought I should cover. Many years later, Enid asked me which part of my career I thought had been the most satisfying? After a little thought I said that my time at Woolley Colliery fitted perfectly, because I was near to the point of production and I could talk to the teams of men and influence them directly.

While I was Deputy Manager, another significant industrial relations case arose. In the 1950s and 1960s absenteeism was very high. On Mondays and Fridays there could be up to a quarter of the manpower missing. A workman, nicknamed Bubbles, was very good at absenteeism. The trouble was that Bubbles had an alibi that protected him. His wife was showing favours to the NUM Secretary at the mine. While he remained on the union, action against Bubbles was not practical. But he retired, and there was another purge on absenteeism. The procedure was followed, and Bubbles was in the first batch to be dismissed, and the union was consulted on the actions to be taken. On the proposal to dismiss Bubbles their reaction was supportive of management: 'Serves him right. He's a lazy bugger.' So, said Elijah Benn, end of case.

But it was not the end of the case. Bubbles' wife went to see Elijah Benn's wife. 'They are going to sack Bubbles,' she said.

'What about the three kids? What am I going to do? And the union won't fight the case.'

Elijah Benn's wife challenged her husband. Case reopened. It was not about Bubbles, it was about three kids who happened to have a lazy father.

Elijah put it simply, 'I am not arguing for Bubbles. I am arguing for them three bairns to have a chance in life.'

The whole pit was watching for the outcome of this case and it went to Calvin Round to arbitrate on the matter.

It was late one Friday afternoon that I came back onto the case. I had been delayed underground at the pit, sorting out a breakdown on some equipment for which I had specialist expertise. I was having a shower when Calvin came to talk to me.

'What are we going to do about Bubbles? Elijah is hanging in and won't let go,' said Calvin.

I was adamant. 'If we concede on Bubbles' case we have no future tackling absenteeism at the pit.'

Calvin supported me. 'Right, that's it then.'

He went back in to see Elijah and told him that he would not concede. Elijah put on his flat cap and said 'Right, Mister. I hear what you say. I've done my best. That's it.'

Bubbles was sacked.

There was a postscript to the story. Bubbles got a job with Shaw Carpets in Darton, and became a regular and reliable worker. The family were better off than they had ever been. Maybe Bubbles was never meant to be a miner. There is a moral behind this case: in management you have to do what is right. There are often factors outside your control that make the consequences of difficult decisions very different to what is expected.

A regrouping of the pits in Yorkshire led to Calvin Round being transferred to Grimethorpe Colliery as General Manager and I was left in charge at Woolley until the position became clear. It was an exciting time as it was near the financial year end and it was possible for the pit to exceed one million tons for its annual saleable output, and break even financially. There was a very positive approach throughout the pit for the short time I was in charge, and both objectives were achieved. It was then decided that the Colliery General Manager post should be filled by a Scottish mining engineer, George Duncan, whose pit had been closed by an underground fire. This released me to follow Calvin Round to be Deputy Manager at Grimethorpe.

It would be nice to report that I had left Woolley set up for its future, however, there was a technical problem in the Lidgett seam. The limited seam height suggested that operations would be easier if the armoured face conveyors could be reduced from seven inches high to five inches high. The change would also allow each web mined by the coal cutting machines to increase in width, giving more output. Five-inch conveyors were installed, but to facilitate the thinner chains on the conveyors the steel was hardened to transmit the horsepower required. The chains were too brittle and broke quite frequently. One of the early decisions by my successor was to change the Lidgett coalface conveyers back to the standard seven-inch-high line pans and thicker but less hard chains.

My last few days at the pit were spent in limbo, as my presence was not required at the routine meetings for the pit as I would not be involved in carrying out the decisions made. This was the loneliest time in my whole career.

### *Deputy Manager, Grimethorpe*

I took up my post at Grimethorpe in 1968, but the family remained living at Woolley, so our whole life became slightly surreal. The Deputy Manager appointment was to last only for a year, and during that time I had a spell in hospital. Grimethorpe was a complex organisation, as evidenced by the number of different activities on the site. The pit had three shafts, two winding coal from different Grimethorpe seams and one winding the output from Houghton Main colliery, which was connected underground. The output from both pits was treated in the Grimethorpe coal preparation plant which produced various grades of coking and industrial coals. (At that time the rumour was that the large, lumpy, best house coal was used at Buckingham Palace.) The coking coal was passed over to a separate plant which produced coke for the steel industry. There was also another colliery on the site, Ferrymoor colliery, which worked the seams near the surface and had its own coal preparation plant. In addition, there was a pulverised fuel plant and a large brickwork, which was still in operation in 2017. If there were any incidents on the surface, the first task was to find out the exact position of the incident to see which authority was responsible for handling the matter.

One memorable underground event in my time at Grimethorpe related to the Parkgate seam, which was over 600 yards deep and had very hot working conditions. A new type of oil lamp, the Garforth lamp, was regarded as a significant technical development for mine deputies to use, as it displayed the methane concentration in the airstream to much more accurate levels than the previous oil lamps. On its first day in use, a deputy rang out of the pit reporting that there was twelve percent methane content in the return airway of 5's coal face. The initial reaction was utter disbelief, but in fact the report was correct. A major outburst of methane had come out of the shale beds beneath the seam on the coal face. A significant programme of work was organised to tap into the shale bed and get the methane into an eight-inch diameter pipe range and bring it out of the pit where it could be utilised. Having experienced this incident, with others later, it appears to me that the pressure groups campaigning against fracking for shale gas, and doubting its availability, may be guilty of disingenuousness.

My period at Grimethorpe also introduced me to the famous Grimethorpe Colliery band. At that time the band was conducted by George Thompson, and all the players were on the colliery books. Some of the players were qualified to work underground and one of my responsibilities was to organise the work they did when they had shifts underground. I saw them on some occasions stripped to the waist on specialist jobs and their pay on those shifts was related to the work involved. Their main preoccupation was to rehearse for their contest and concert work, which took at least two days per week, so their input to the colliery performance was minimal. Their performance as musicians was magical, though, and they were one of the finest brass bands in the country. The band is still in existence, and celebrated its centenary in 2017, playing brilliantly but struggling financially. On one occasion I was undertaking a surface inspection with the surface foreman and we were discussing a certain operation. The foreman had with him a surface worker who was dressed in dirty, rough clothes and a flat cap. The worker looked faintly familiar and I suddenly realised that he was the lead cornet player of the band, namely Brian Cooling. What a contrast for him — his job on the surface at the pit and then standing in evening dress playing for the band on the stage at the Royal Albert Hall!

There are occasions when good fortune is abundant and one such was when the engineers were changing the guide ropes in one of the shafts. These ropes go through hoops on the cages or skips and are held in tension at the bottom of the shafts with weights. They ensure that the cages or skips don't collide as they pass each other in the shaft, and also align the cages so that they land at the correct position at the surface and in the pit bottom. They are large diameter ropes and, at 600 yards in length, they weigh a significant tonnage. When one of the ropes was being lowered into the shaft the capel holding a rope on the reel at the surface fractured and released the rope, sending it cascading down the shaft into a coil in the shaft bottom. There were men in the pit bottom, and one or two actually in the shaft itself monitoring operations, and not one was touched by the rope as it rapidly snaked its way through the shaft. It could have been an accident with multiple deaths.

Grimethorpe had one of the best First Aid teams in the industry. They were involved when a major roof-fall, at a junction where men were working, resulted in fatalities. I was away on holiday at the time, but I learned of the incident from Calvin Round, who had visited the site with the mines inspector. The First Aid team persuaded the manager and mines inspector to move away from the site and they then managed to rescue a couple of men buried under the fall. With the Grimethorpe NUM secretary, I visited one of these men in hospital. His injuries were significant but not life threatening. However, he was clearly suffering from major fears and shock from being buried. He needed lots of support from his workmates and his family before he dared to return to his job.

# Chapter 12

*A new mine and a big mine*
*1969-1972*

At the end of one year at Grimethorpe, a new mine concept was nearing completion in the Area and it needed a colliery manager. I was given two conflicting pieces of advice about the post. 'You must apply for it; it will be the best colliery manager's job in Yorkshire' and 'You must not touch that vacancy with a barge pole, whatever you do; it will be a highly political appointment and a very risky career prospect for whoever gets it'!

There was no compromise between those two bits of advice. Nevertheless, I applied for the post. It was like the Miss World competition in that the man behind the project, Charles Round, interviewed about eight applicants and then had further interviews for three from a shortlist. Fortunately I was successful and got the appointment. Later I was to find out the politics of the project when I took up the post.

There were three key people on mining and business issues in the Barnsley Area at that time: the Area General Manager, Eddie Hoyle, who had held key roles in mine management in pits around Yorkshire; the Area Production Manager, Charles Round, who had held management roles in Yorkshire and in South Wales and had successfully worked short coalfaces at Elsecar colliery in Yorkshire and visited the USA and seen their 'in seam' mining operations; and Bernard Goddard who was Chief Mining Engineer in charge of the mining systems and safety and who also had experience in managing Yorkshire mines. Eddie Hoyle and Bernard Goddard had been colleagues for many years. Charles Round had pushed the project for the new mine at Coal Board headquarters in London, with the objective of combining American continuous miners to drive roadways blocking out the coal which would then be mined using retreat coal faces with UK mining equipment. He was a dedicated mining engineer who worked hard and was forceful in pushing the new technology that he favoured. He wasn't keen on collaboration with Eddie Hoyle and Bernard Goddard; his ambition was to replace them and take over as Area General Manager.

The project for the mine, named Riddings Drift, was simple. In a part of the South Kirkby Colliery surface, an inclined roadway (surface drift) had been driven down to the Shafton Seam and connected through to the Ferrymoor mine workings in the Shafton seam. A conveyor (cable belt) was installed and the Ferrymoor coal was transported to the surface. On additional conveyors, on the surface, it was elevated to a spiral stock pile from where it was loaded into lorries and taken to Ferrymoor pit for washing. There was a man riding haulage down the drift to take the men and materials for the Riddings operations. There was a block of two semi-detached houses that were to be used as offices. There were also small workshops for the engineering operations and a stockyard for holding the mining materials. The stores were managed by the storekeeper at South Kirkby colliery.

When I arrived, the mine might as well have been in the wilds of the USA mining coalfields in West Virginia. There were no management staff appointed. The office block had no telephones, or office furniture. I had a small number of men and four underground officials. The first month of my management was taken up with the important appointments of a management team, equipping the workshops and offices and then establishing the objective of exploiting a new system of mining with limited manpower. The project brief indicated an output per man-shift of four times the national average at that time.

In recruiting staff and workman there were moments of unique challenge. One lady applied for the post of manager's PA. She was an impressive-looking woman in her late-twenties, with a distinctive figure. How good she might have been at shorthand and typing I don't recall because, when answering my questions, she continually crossed and uncrossed her legs. I decided that she would be too big a distraction to anyone wanting to see me, so I chose a young man of limited physical attraction who proved excellent as a PA and he took on all the jobs necessary in an efficient organisation, such as transport, visitors, trainees and supplies.

Wombwell Main Colliery, about seven miles from Riddings, was closing, and their manpower was available for transfer as we built up the operations. The pit was a source of experienced workmen who would need transport to the mine. My Safety/Training Officer had been at Wombwell Main, so I had him in with me when I was interviewing for workmen. One Saturday morning a man who had been a pump man at Wombwell Main pit came in. Sid Wilson was in his fifties, wearing glasses and appeared an old man, not particularly active. My reaction was that we did not need a pump man as our pumps would be automated. After a while, Sid obviously reckoned that I was not going to offer him a job, so he said, 'Mr Massey, if you will give me a job I will do anything you ask me to do. You can send for me anytime, day or night, and I will come, and I will never, ever let you down.' What an offer. I took him at his word and he never failed me.

The management team was slim: one undermanager, one mechanical engineer, one electrical engineer, one surveyor and one surface foreman. They became a magnificent team. None of us were ever office-bound. The action was underground, so we were all underground each day covering various parts of the mine. Two special appointments required extra elaboration. The surface foreman had to supervise the transport of all the materials underground on the man-riding train. As we increased the momentum underground this became a major challenge. Later, when the coal faces were in full production, disposing of the coal from the site became a problem too. On an afternoon the haulage contractor had to call on help from his other contracts to clear the stockpile. The surveyor had to provide guidance for all the roadway drivages, so that they went exactly where they were planned. He did this by establishing theodolite bases along each roadway correlated back to a surface station. When the roadways for each coal face were connected we could rely on the length of the coal face being exactly 60 yards, plus or minus two inches.

The roadways were driven virtually in-seam, with an extraction of about five feet. The roof supports were 14 feet long, 4-inch by 4-inch H-section, girders with wooden props at each end.

The Lee Norse miners extracted the coal and a little floor dirt. They were track-mounted, with an elevating frame at the front on which was mounted two drums laced with picks which cut the coal. The drums rotated and oscillated, starting to cut at the roof and then ripping off the coal down to the floor level. A flexible chain conveyor through the centre of the miner transported the coal and dirt onto a belt conveyor. Each side of the roadway was advanced by three feet and then a roof support was set. On the front frame of the Lee Norse was a bracket used to lift the girder up to the roof and hold it in place while it was clamped to the previous support and the wooden props were set. There was a team of six men in each heading and they had to carry out all the work of cutting the coal, setting the supports, extending the conveyor and the haulage system and getting all the supports up to the machine. Periodically there was also the need to advance the electrical cables up the roadway and move up the switchgear. A par performance was ten girders in a shift, but on occasions a team would set fifteen girders.

As the workings were extended, three continuous miners were deployed and the first block of coal was around 800 yards long by 650 yards wide. It contained ten coal faces, 60 yards long, which each had 600 yards to retreat from its start point to its finishing line.

The coal face system was to use a ranging drum shearer with a 30-inch diameter drum which cut the top coal as the machine travelled from the main roadway to the return roadway. At the end of the run, the machine parked across the roadway and a cowl around the drum was pulled over and the drum lowered to cut up the bottom coal. On the return run, the panzer conveyor was rammed over and the powered supports set ready for the next shear. There were only five men in the team, with a deputy in charge, and they developed a rhythmical method of work which fascinated visitors, including mines inspectors and senior Coal Board officials. The men never appeared to be rushing to do their work and they all knew exactly when to play their part. The result was a steady, almost continuous, stream of coal coming from the coal face. Even on the first coal face it became clear that we had established an efficient mining system.

The challenge for the management team was to achieve levels of performance from the continuous miners so that coal faces were opened out ready to replace the ones that had completed their run. Initially we only had one set of coal face equipment so when a coal face finished, all the equipment had to be transferred to the new face.

*Alf Robens underground at Riddings Drift*

The surveyor drew up a large plan on the wall of his office which showed how much coal we had in the bank from the roadway drivages at any time.

It became the policy that nothing should stop the three continuous miners. We developed plans that required continuous miners to cross over each other's roadways rather than turn off junctions which delayed operations. It also became clear that a critical job was moving equipment around the mine — conveyors, fans and electrical gear. A very experienced team of men in the pit were put onto this role. They were divided into two groups and worked on split shifts — 8am in the morning and 6pm at night — so that they could move equipment outside the main production shifts.

There was also the incentive to simplify systems. Initially there was a stage loader which received the coal from the face conveyor and delivered it to the roadway conveyor via an overlap. This meant that the production stopped when the roadway conveyor had to be shortened. We reduced the stage loader in length and coupled it directly to the tail end of the roadway conveyor. When the face advanced, all that was required was to tighten the conveyor belt in its loop and take out sections of the conveyor structure as necessary. I recall being severely criticised by a senior official for altering a successful system. However the change worked, and the coal faces became even more successful. There was also a change in the surface arrangements as the output grew. As the coal arrived at the surface, the run of mine coal was passed over long screens which scalped off the minus-one-inch coal. This was transferred to South Kirkby colliery and went into the power station coal. As we were working in the seam the coal did not require an expensive washing process.

It became clear that the potential of this mining system was huge. This led to a second set of coal face equipment being provided so that there was continuous output from a production face. The operations on the coal faces settled on a performance of two shears per hour, which gave good levels of output per man-shift and overall profitability for the mine.

### *Official opening ceremony of Riddings Drift mine*

Eddie Hoyle, the Area Director, led the initiative to have an official opening ceremony for the mine. Alf Robens was the Chairman of the NCB and he had seen continuous miners at work in America, but had never seen one in use in a British mine. The day was to start with an underground visit on the Saturday morning, for Alf Robens and senior Coal Board and Area staff, who would visit the coal face and one of the continuous miner drivages which was operational. After the visit there would be a lunch in marquees set up on a surface area adjacent to the mine. Every man on the books — at that time about 180 — would be invited, along with their wives. Some wives were not welcomed by their husbands, as they considered it inappropriate for women to be at such an event. It was a large marquee and, in an extension, Grimethorpe Band played during the meal. (A young member of the band had composed a special piece for the event and, when the band had performed it, Alf Robens went over and complemented the composer.)

Organising the underground visit had its problems, as the coal face had a fault on it which produced some water on the face. We had to lay special duckboards so that the visitors could get to the coal face without getting their feet wet. The height of the roadways, at around five feet, were an awkward height for walking and you had to adopt a crouching pose with bent knees to move without banging your head on the roof girders. So, we had to improvise a special man-riding system to transport the VIPs around the mine. The many hours put in by the management team had paid off and the underground visit was a success. Some excellent photographs were taken underground of Alf Robens in discussions with the workmen.

After he had showered, in the South Kirkby manager's office, I accompanied Lord Robens up the road to the marquees. There, at the entrance, I saw a crowd of pickets, with banners and placards, complaining about housing repairs in the local NCB houses. I was stunned. After all our efforts, the day was to be ruined by these pickets. My concerns must have been apparent to Alf Robens. 'Just leave this to me,' he whispered.

At the entrance to the marquees was a catering company waitress, in all her finery, offering glasses of sherry to the guests. Alf called her over.

'Please give these good people here one of your drinks.' Alf gave the pickets a smile as they accepted the offer. They took the drinks and melted away.

After the meal, Alf spoke to the assembled workers. Whilst he had no notes, he was inspirational in his comments to the workmen and the management. Every person left that ceremony motivated to be a superman. They couldn't wait to get back to work. I had the job of responding. I can't remember exactly what I said, but I tried to assure him that we would make a success of the mining opportunities.

Shortly after this visit I departed to the USA for four weeks, visiting mines across the country. Two of us had been granted the Tom Seaman Travelling Scholarship offered by the Institution of Mining Engineers in 1970. It was an intensive programme and left a lasting impression on both of us. Of the many incidents we witnessed I will mention just two. We visited Peabody No.10 mine, which was the largest underground operation in the USA, at the time. The Vice President of the Peabody Company, Joe Craggs, spent some time with us and he invited us to have dinner with him one night at his house. He said he would also invite his mother along. The house was quite magnificent, with a great deal of land — almost like a ranch — and outside it was parked a mobile home the size of single-decker bus. His wife, a charming and devoutly religious lady, had an electronic organ in a large room. Joe's mother was a little old lady who had travelled to the USA on a cattle boat fifty years previously. Remarkably, she still spoke with a Durham accent, as she originated from Trimdon Grange. This family represented to us an example of the opportunity to succeed in the USA. To go in one generation from landing in a cattle boat to becoming a millionaire and a very successful senior manager in a competitive industry, was quite something.

My second example comes from the coal face operations that we saw at several mines. They were in the initial stages of using UK equipment, and trying to operate longwall coal faces. Their coal face shearers were identical to the ones we were using at Riddings Drift. I was surprised at the rate their machines were producing coal. The seam characteristics seemed similar to our UK coal seams, but the rate of production was obviously higher. I enquired about this and was advised: 'If the disc on the side of the cutting machine says 200 horsepower, we make sure that 200 horsepower is going into those cutter picks.' I needed to follow this up when I got back to Yorkshire.

Back at Riddings, I asked the electrical engineer what horsepower our shearer was using. He replied that it was operating well below its rated capacity. Why, I asked? The power was controlled by a circuit in the haulage section of the machine. Above the set figure it cut out the pull by the machine on the haulage chain, reducing the applied horsepower. So we discussed this with the mechanical engineer. What was the cut-off pressure in the haulage end? About 2,400 psi. Was there any reason why it should not be raised marginally? An increase of 200 psi was requested. The result was to double the production of the shearer. My aim had been to increase the performance to three shears per hour from two per hour. It actually became possible to achieve four shears per hour when all the team were on song. This meant that when the coal face was operating I could go to bed on a Tuesday night and know that every ton of coal from then on through that week was pure profit. Most of the record performances at Riddings were achieved after I left the pit, but one I retain.

When the Production Manager, John Williams, called me one morning, in his usual style he asked. 'How did you get on yesterday, young fellow?'

I replied, 'We did 72 shears, Mr Williams.' For the first time ever in our relationship he was speechless. My personal objective was to produce more coal in one week, with my 180 men, than South Kirkby colliery did with its two thousand men. We achieved that objective. In the first year of operation, the mine achieved a profit higher than the capital investment to develop the mine!

In the NCB at that time there was a clear division of views about retreat mining. In the Nottingham Area they favoured advancing coal faces, because they were used to them. My view was that driving the roadways at the face ends interfered with the coal face operations and they advanced about half the rate of driving independent roadways. There was another advantage in retreat systems: the developments for the retreat faces found out about the geology of that area. There were cases when unknown faults stopped advancing coal faces and seriously affect the economics of the mine. I accepted that the ventilation of retreat faces needed to ensure that the methane levels in the waste behind the coalface had to be controlled, but techniques to do this were evolving, even for mines which had high methane levels.

The success of the Riddings Drift was not met with universal acclaim. I was at pains to try to avoid the politics generated by Riddings within the Area. When Charles Round had been on holiday for a week, before he went back to his office at Area, he would come to my office for an update. 'Who has been while I was away, and what did you tell them?' Finally, Charles lost his battle and Eddie Hoyle had him removed from his position, to a job at Headquarters. His place was taken by Mike Eaton, who was slightly younger than me, and we were to work together, on and off, throughout the future.

### *General Manager, South Kirkby Colliery, 1972-1973*

Having produced more coal at Riddings with less than 200 men compared to the 2,000 men at South Kirkby Colliery, I suppose it was to be expected that someone might want to try me as General Manager at the big pit. Mike Eaton moved me across. So, when I accepted the offer, I wrote a letter to the staff and officials at Riddings, thanking them for their hard work and support during my time as manager. Whilst dictating this to a charming young lady who worked in the Riddings surface office, tears were streaming down her face when the contents of the letter sunk in, so I had to excuse her and write it out longhand for her to type!

Managing a large pit, with 2,000 men and working three seams, was a major change and a challenge after managing Riddings Drift. Fortunately, my style of management was very different to that of the person I followed. The men noted the differences. 'He gets down the pit regular. He talks to the men. He brings sandwiches and doesn't go out for lunch.'

It was necessary to get the management, men and union officials to believe that the pit could be successful. I set up a regular meeting, every Monday lunchtime, with the workmen and officials from some part of the pit, along with the union representatives. Some issues were raised that would increase performance and some matters were raised that would irritate the men. It was important to demonstrate that I was prepared to listen and make changes.

In any new management situation, where the prospects are recognised to be poor, it is important to be a sponge, seeing all aspects of the operations, ignoring the opinions of those steeped in failure, and taking into account glimmerings of any opportunity for change. After I had been at the mine a few months, my hopes became more positive. One night I said to Enid, 'The problems at South Kirkby might not be as hopeless as some people think.'

Vince Hollingworth, the Deputy Manager, was a key member of staff at the pit. He was a very hard-working and respected mining engineer, who could always be relied upon for his assessments of people and operational problems. He was a fanatical cricket fan and still played, in his forties, for the Hemsworth team. His speech was often peppered with cricketing terms: if something was successful he might preface his remarks with, 'We're going to score sixes with that set up, Boss'; if it was negative, he would look depressed and say, 'They've bowled us a right googly on that job.'

The main problem was to get the strategy right for the pit. An entrance had been made into an area of the Barnsley seam which was ten feet thick. Some people, including Johnny Williams, the Production Manager, decided that this should be the base load for the pit. Working that seam, with retreat coal faces, presented special ventilation and spontaneous combustion challenges that had to be solved. In the Meltonfield seam there were three coal faces, with excellent roof conditions, but the powered supports were useless and leaking fluid all over the place, so that the men were crawling about in mud and water. And then the pit faced the 1972 miners' strike — a time to take stock.

The 1972 strike was a challenge to the Coal Board, as it was the first national strike in the coal industry since 1926. The layout of South Kirkby pit was ideal for the NUM as there was a single, narrow road access to the pit. The NUM set up a picket along that road and stopped every vehicle for inspection. Management were allowed to go through, but the NACODS officials refused to pass through the picket line. So, the management team at the pit had to cover the mine for safety inspections and engineering maintenance. The management team were deployed throughout the pit doing remedial work to preserve the mine for after the strike.

The NACODS officials still wanted their weekly pay, but they were the wrong side of the picket lines. The prospect of them being allowed to pass through the picket lines to get to the pit to collect their pay, and then returning, with their pockets full of money, back through the picket lines, seemed a scenario for trouble. So special arrangements were made to use another pay point, at a site where there were no picket lines. The particular site was only notified to the NACODS members on each pay day.

We had stacks of different coal grades around the mine site. One stock of best house coal was near the perimeter of the site and we realised that this was being pilfered as there was a good market for it. We decided that it should be moved to a site near the colliery offices, where it could more easily be observed. The chief engineer operated a pay loader, I drove a large dump truck and Joe Cooper, the colliery Admin officer, acted as signaller to guide me to tip the coal neatly on the new site.

It was during the 1972 coal strike that I rang Mike Eaton and convinced him that a major change in the pit's performance would result if the powered supports on the Meltonfield coal faces were changed. He agreed, and two new sets of Dowty powered supports were delivered during the strike, with a third set to follow. After they were installed the output from the Meltonfield seam increased dramatically.

Following the strike, the solution to the ventilation in the Barnsley retreat coal faces was solved, by special arrangements ensuring that the methane was cleared from the coal face operations. To demonstrate that this was happening we instituted a continuous sampling system at specific places along the coal face, taking hundreds of samples each day, which were recorded. This involved work with the mines inspector so that he, along with our management team, was satisfied that we had a safe working practice.

An interesting situation arose at the Miners Welfare Clubs during the strike. I had two under my jurisdiction, and they were chaired by my Deputy Manager (Hemsworth), the Assistant Manager Personnel (South Kirkby); I attended meetings when there were major issues to solve. I was asked by Vince Hollingworth to attend a Hemsworth meeting when it was rumoured that a motion would be tabled to reduce the price of beer in the Club, to help the picketing miners. When it was raised, I spoke against it, saying that the Club's assets, which were significant at over £20,000 invested, should not be wasted. We didn't know how long the strike would last, and we might need the assets for food kitchens and to feed destitute children. The motion was withdrawn. A week later a motion 'to take up Mr Massey's suggestion to give food parcels to all the families who had a member on strike' was agreed. The committee worked hard preparing and distributing the substantial food parcels. The community recognised that this policy was far better than selling cheap beer and, as a result, the beer sales at the Club stayed at the pre-strike level and the food parcels were financed without touching any of the significant financial reserves of the Club.

Towards the end of the strike I was able to do an additional service to the mining communities. I had a bunker in the coal preparation plant that was full of coal suitable for using in the boilers at local schools who were facing closure without any heating. I got permission to send coal to several schools and also to the two Miners Welfares, again getting approval from the communities.

The Wilberforce enquiry, which formed the basis of a settlement of the 1972 strike, lifted the miners' wages significantly, but high levels of inflation soon wore down the differentials. This led to the national three-day week, and another strike in 1974. The government was closely involved in the negotiations and an election was called on 'who ran the country'. The result was the change to a Labour government which splashed out money on many aspects of pay, such that the seeds of the end of a viable coal industry became a possibility.

Two other significant matters arose while I was at South Kirkby. One morning, when I was in my office, I was asked to take a call from Brynnen, my eldest son. I suspected that there must be some problem as this was most unusual. However it was good news, and he told me that he had obtained a place at the Queen Elizabeth Grammar School in Wakefield. I was delighted, and congratulated him. The reason for him trying to go to that school arose from experience with the local schools around Hemsworth. When we lived at Woolley Colliery, the village school had a brilliant headmaster who inspired the pupils and introduced them to many extra-mural activities, as well a sound teaching in academic subjects. But it was a small school, and Enid considered that our three children needed the additional competition which would come in a larger school. When we moved to Hemsworth, the local authority had decided to introduce middle schools between the junior schools and the grammar schools. Teachers were against the change and there was uncertainty about the educational prospects.

So Brynnen took the entrance exam at the Wakefield Grammar School, which was my old school. When I got home that night and we were discussing Bryn's success, Enid set out our future policy.

'You realise that if Bryn goes to QEGS, the other two will have to go as well.'

There are only three and a half years between our three children, and the financial implications were significant.

'You must be joking,' was my reaction.

However, Enid's plan became a reality, and Adrian and Dawn also went to the Wakefield schools. We believe it was the starting point of three motivated students, working hard at school and university, gaining good degrees and successful careers. Fortunately, my own career had a continuous flow of promotions and increases in salary such that we could afford to have one third of our net income directed into the children's education.

The second significant matter is more trivial, but nonetheless relevant. Joe Cooper was the Administrative Officer for South Kirkby. He was 'Mr Fix-it' and his answer to any question was always, 'No problem, Boss. Leave it to me, I'll sort it.' I'd changed my car, which proved helpful crossing the picket lines, as the pickets didn't recognise my new car!

After the strike, as I was going underground one morning, I said to Joe, 'If you know anyone who wants a car, my old one is for sale.' I gave him an idea of the price and he replied, as usual, 'Leave it with me, Boss.'

About two hours later, I got a message underground that Joe wanted a word with me. I rang his office and got the message, 'I've sold your car, Boss.'

Later, I found out how he had managed it. A young management trainee had visited Joe in his office with a request. 'Will you have a word with the Boss, Joe, and ask him if he will agree to me having a few days off so that I can repair my car which has packed up?'

Joe looked at the young man and offered him some advice. 'If you are going to make good in management you must have a reliable car; it is essential. I know where there is a very good reliable car, at a reasonable price.'

To conclude the sale, Joe arranged to transfer the young man's bank account from the Barnsley branch to the Hemsworth branch of the Midland Bank. Joe then had a word with the Hemsworth bank manager and agreed a loan to finance the purchase of the car. There is a role for a Mr Fix-it in every organisation!

I knew that the pit was on target to achieve much improved performances. When I arrived, the pit struggled to achieve 15,000 tons per week. When the new supports were installed in the Meltonfield seam, and the Barnsley seam was in production, I predicted that the pit could achieve 30,000 tons per week. It did, after I had moved on, but it was a thrill for me to see the management team revelling in the high output and good financial results for the pit. I was succeeded by Roy Beeforth, who was my undermanager at Riddings Drift, and had taken over my manager's role there. I knew South Kirkby was in good hands.

## A musical interlude

Within the Coal industry there were many opportunities for taking part in musical activities. I have already referred to brass bands. Male voice choirs were another example. My Dad was a good tenor singer, and after he finished with the Methodist church choir in Higham, he joined a male voice choir which was formed at Dodworth Colliery. He took me along with him before I went to university. This led to my sudden elevation to conducting.

The choir was entered for a competition, and they had a very good conductor preparing them for the event. At the final rehearsal, though, he sent a note informing the choir that he could not continue. He was a quiet man who probably lacked confidence for major performances. What options were there for the choir? However, they decided that, because I could count, I should conduct them at the contest! They had been well prepared, and I encouraged them to be confident. They won!

I conducted the choir for a period, but promotions in my career meant that I couldn't continue. However I learned that persuading a choir to jointly combine to make an attractive sound is as much a psychological challenge as a musical one. The Coal Board organised the Yorkshire Coalfield massed choirs and brass band concerts, held in the City Hall in Sheffield and conducted by Reg Jenkins. I was able to sing in a couple of those.

Then, eleven years later, my Dad joined a choir in Barnsley which rehearsed in the Huddersfield Road Methodist Church. They had a mixture of experienced singers and young learners. They asked me to conduct them and I took up the challenge. Their aim was to perform concerts, but they didn't want to spend long periods rehearsing for competitions. This led to a development that gave some outstanding concert performances.

Margaret Firth, the daughter of one of our bass singers, had a beautiful, powerful soprano voice. She appeared in solo lead roles for the amateur operatic society in Barnsley. We approached Margaret for her to sing solos and some items with the choir. I recall that when she first came to one of our rehearsals, she stood at the side of the piano to run through one of her solos. As soon as she sang the first notes, there was a gasp from the choir members. She was something special. What impact might she have on the choir? But we had another ace card in our hand. One of our members was a piano tuner and an organist. If we were giving a concert in a church, we sent him ahead to tune the piano to the same pitch as the organ. One item in our programme was 'The Holy City.' We adapted the music so that Margaret sang the verses and the choir sang the choruses, but for the final chorus the organ joined in and Margaret and the choir sang that chorus. We altered the last two notes, so that Margaret sustained high notes on the word 'ever', but the tenors sang a glorious final high note for the last word 'more'.

Numerous people told me that when the organ joined in, shivers ran down their spines. The choir were inspired, by the audience and by Margaret, to sing their hearts out with that final chorus. It was always a glorious finale.

This showed that amateur musicians were able to create moments of magic — when the right balance of musical skills were brought together at the same time.

In October 1973 my father died, after suffering with stomach cancer. He was 65 years old. His suffering through that year was a great strain on my mother and the rest of the family. At his funeral service, the choir swelled the singing and showed their respect for him. Without seeing his face in the choir as I was conducting, my future role with the choir was just too much for me to bear and so that was the end of my conducting career.

# Chapter 13

*Production Manager, South Yorkshire Area*
*1973-1975*

My next move resulted from changes in senior management in Yorkshire. Eddie Hoyle had moved to be Area General Manager in South Yorkshire Area, Mike Eaton took his job as Area General Manager in Barnsley Area and Albert Tuke became Area Production Manager in Barnsley Area.

Albert had made his name as manager of Silverwood Colliery in South Yorkshire Area, where he established intensive operations on one coal face at the pit. I went to visit it when I was at Grimethorpe colliery. He set up the coal face to operate to a rigid timetable, like a bus service. On this basis they did six shears in the one shift. They worked the roadways and face ends on three shifts to keep up to the programme. It was most impressive. His post as Production Manager for the Manvers group of pits in South Yorkshire was on offer. I applied, as a speculative promotion, as there would be a major benefit: my location in Hemsworth was right in the middle of the Yorkshire coalfield. I could work in any part of the Yorkshire coalfield without relocating. This meant that the children had continuity of their education.

The role of Production Manager had evolved from the former post of Group Manager, when a mining engineer had responsibility for a group of collieries. The Group Manager was supported by a small team of specialists, which in some cases was inefficient. Specialist engineers were now held in the area team to support all the pits and were on call to the Production Manager as required. The Production Manager's role was to support the Managers and visit and inspect the pits. Each mine had a detailed action programme showing which coal faces would be in production at any time in the next three years and the development work to get them ready. There were also capital programmes at each mine. The Production Manager held meetings each quarter with the Managers, to review the pit's action programmes.

The Area Director held an accountability meeting each quarter with the Colliery Managers and their Production Managers to review the financial performance of the mine against its budget. If the Colliery Manager was under pressure because his results were not up to his budget figures, and he was hesitant in replying to questions, it was tempting for the Production Manager to join in. Eddie Hoyle was fiercely against this, and on one or two occasions I was given a stern look and a pointed finger. It was the Manager who was being held to account!

The interview for the post was conducted by Eddie Hoyle, with Charles Dickens the Area Production Manager and the staff manager. I assumed that there would be some candidate within the South Yorkshire area who would be favourite for the promotion. That was true, but I was appointed, probably because of my successful experience with retreat mining systems.

The South Yorkshire Area had 19 pits and produced high quality coal, mainly for the coking market and it was highly profitable. The Manvers Group was made up of four pits — Manvers, Wath Main, Barnburgh and Kilnhurst, which operated as a central combined scheme installed in the early sixties. It was a very successful project. The coal from Manvers, Wath and Kilnhurst pits was brought to the surface through two dedicated shafts at Manvers. The coal from Wath and Kilnhurst was transported in seven-ton mine cars and hauled underground by diesel locomotives to the two pit bottoms at Manvers. The mine cars were emptied in the pit bottom and the coal transferred to skips to be wound up to the surface. The upper seam Manvers coal and the Wath coal was wound up the N°2 shaft and the Manvers Silkstone seam coal and Kilnhurst Swallowood seam coal was wound up N°3 shaft. The coal from Barnburgh pit was brought to the surface at Barnburgh and transferred in railway wagons overland to the Manvers site.

All the coal was fed into a very large central coal preparation plant and treated to make a range of products. The main product was coking coal, which was treated in a coking plant on site. The coal preparation plant consisted of three dense medium units that treated the large coal. There were then three Baum boxes that processed the minus-three-inch size coal. The fines were processed in filtration units to separate the very fine coal from the fine shale and waste. All the dirt was transported onto a large tip about a mile from the central site. It also included a large lagoon to take the waste water from the site in the middle of the tip. This was a potential hazard, as there was a public footpath passing near the tip. The coal from each mine was treated separately and, at the end of the week, the total saleable coal was apportioned back to each pit. The logic of this share-out, made by the Manager of the Central Coal Preparation Plant, was not clear to me. It was rumoured that he was influenced by which colliery manager had bribed him with free golf balls!

The person who was favourite for the Production Manager's appointment was George Hopcraft, who was Manager of Barnburgh pit. He was a tough manager, who worked very hard with some difficult union men, but, when I arrived, his pit was the least successful of the four. It is interesting to record that in the final analysis, much later, his was the last of the four pits to be closed. I got on well with him, and we were able to introduce some retreat working at the pit. I was mystified as to why his share-out of the saleable coal from the central coal preparation plant seemed so low. In travelling around the pits, my impression was that the Barnburgh mine seemed to be providing the cleanest run of mine coal to the plant. I introduced sampling of the run-of-mine feed to the Baum boxes for the different mines, and the share-out of the saleable coal was based on the quality of those run-of-mine samples. Barnburgh benefited, while Manvers had a loss.

I learned, after I had left the South Yorkshire Area, that George had said that he thought my appointment was justified as I had introduced retreat mining to the area.

Manvers was managed by an inspirational character with a fine mind — Richard Brooks — who, at his best, could lead his men to improved performances. However, he was affected by his dependence on alcohol. It presented some tricky situations for me, and I depended on the Deputy Manager at Manvers, Peter Lawrence, to keep a watch on his manager and avoid any unacceptable behaviour. (Peter Lawrence later became Manager of Silverwood colliery. He took the Queen underground on a visit and there was a wonderful photograph of her smiling at some remark as she waited for the Duke on the man-riding train. That photograph was displayed at every site of the NCB.)

The coal faces in the Group were mechanised, mostly with powered supports. The faces in the Manvers Silkstone seam and the Kilnhurst Swallowood seam produced coking coal, but the working environment was hot and oppressive.

The Wath mine coal faces had good working conditions and they had evolved a partial retreat system. They worked in blocks of three coal faces. An advancing coal face was worked out to the boundary. While this was taking place, a roadway was driven out 250 yards away on each side of the advancing coal face and these were used as the intake roadways for the flank retreat coal faces. Each roadway of the advancing face was used as the return airway for the retreat coal faces. When the advancing coal face was finished, the supports were salvaged and moved over onto the first retreat face. Thus, Wath had the added efficiency of a proportion of its output being from the special retreat faces. However, the capacity of the shaft winding the coal at Manvers was a limiting factor on the potential at Wath pit.

Wath was a profitable mine; Manvers too normally operated at a profit. The real problem was Kilnhurst. It had a thick seam section and it was good quality coal, but it never seemed to achieve its potential. Accountability meetings for Kilnhurst were never comfortable!

The 1974 miners' strike, and the three-day week, happened while I was a Production Manager. Again, the management had to carry out the safety examinations in the mines. I was informed that in the 1972 strike, the daily shaft examinations had been shared out among management staff, some of whom were terrified of having to ride through a shaft on top of a cage clamped just to the winding rope. This was unacceptable to me on safety grounds. So I told the managers that I would do all the shaft examinations with a deputy mechanical engineer as my assistant. We took over an old Landrover so that we could travel to the different mines in our shaft man's gear. We had a special 'phone that we could use to keep in contact with the onsetter in the pit bottom, and the banksman at the surface, who could signal instructions to the winding engineman, who was also a member of management. I wasn't afraid of the winding engineman making any serious mistakes, as we generally travelled slowly to do the examinations. The plan worked, and we were able to make a positive contribution to the management teams.

One interesting event happened whilst Eddie Hoyle and John Mills, a Member of the Coal Board, were on tour in the area, visiting the pits to support the management effort. They came to the shaft side at Wath when we were preparing to do a shaft examination. We spoke to them and explained that we were aware that there was a leak from a water pipe that was spraying water across the shaft and wetting anyone riding in the cages. We intended to wrap the pipe to stop the spray flying across the shaft. We had a piece of conveyor belting and some wire to do this. We went down and stopped at the leaking pipe and carried out our repair. What we didn't realise was that the water followed the pipe down to the next steel girders across the shaft and then splashed out in all directions. In our exposed position on top of the cage it was like being in a torrential thunderstorm. Our phone ceased to work so we could not communicate to ask the winding engine man to speed up the decent. When we surfaced, we were like drowned rats, which might have generated some sympathy from Eddie Hoyle and John Mills if they had seen us in that state!

There were five Production Managers in the Area, one of whom was Charles Turner, my former mentor. At my first monthly meeting with

Charles Dickens, the Area Production Manager, my colleagues offered me a seat in the centre of the room, directly in the sight line of Charles Dickens when he lifted his eyes up from his papers. They were wily, experienced mining engineers!

Subsequently, Eddie Hoyle took the five Production Managers out to dinner as he was reducing the complement to four Production Managers and putting one on special safety duties. We thought it was a somewhat basic menu as each place was set with just a knife and fork. The menu was Barnsley Chop, and the chef demonstrated how to carve the lamb meat from the bone. It was beautiful, tender meat and a very satisfying meal. I got Cortonwood mine into my group — not a great prize. It was a small mine, working the Silkstone seam and producing coking coal. It offered little scope for technical development. It reached fame later as the mine whose proposed closure sparked the ballot in the Yorkshire coalfield which was used by Arthur Scargill to launch the National Miners' strike in 1984.

It was about this time that the organisation of the areas was re-defined. The head of each area became the Area Director; under him was a Deputy Director Mining (DDM) and a Deputy Director Administration (DDA). The DDM was responsible for the mining operations and the DDA was responsible for all the administrative services within the area.

After less than three years, my enjoyable time in South Yorkshire came to an end with a move to the Doncaster Area.

# Chapter 14

*Doncaster Area, Chief Mining Engineer*
*1975-1977*

The Doncaster Area had ten pits, which were along the eastern flank of the Yorkshire coalfield. Nearly all the workings were in the concealed coalfield, so the mine shafts had to be sunk through the Permian strata, which presented major challenges with water during the shaft sinking process. The workings were deep, but the reserves were extensive, and the coal seams were present through to the east coast and out under the North Sea. (The North Sea oil and gas reserves are below the coal measures in the North Sea.)

The pits had a reputation for militancy and poor labour relations. In some ways the Coal Board was trying to establish a senior management team in the area that would change relationships such that the mines could achieve performances comparable with the deep mines in Nottingham. The Area Director, Jack Wood, who had been very successful in senior management in other coalfields, was a character who revelled in talking to anyone with friendship and humour. Ted Robson, the Deputy Director Mining, was from the Northeast and his aim was to avoid confrontation and apply modern mining technology in the mines. My style of management fitted neatly with both of theirs.

The role of Area Chief Mining Engineer was focussed on the specialist services in the Area. There were engineers in planning, surveying, mechanisation, safety, dust suppression, spontaneous combustion, subsidence, strata control and the underground environment. These were complemented by a range of engineers. The Area Chief Engineer and the Area Chief Mining Engineer were responsible for establishing and monitoring the technical standards for operations in the mines. The basis of these standards was technical instructions and directives issued from time to time from the Mines Inspectorate and from Coal Board Headquarters.

It was a time of major changes in the equipment and mining systems in the industry. There was an elaborate procedure for the field trials of new equipment, with reporting on the trials to Headquarters to gain approval for general use. Bretby was the centre which carried out research into new systems for underground mining. There were two sites, one at Bretby and the other nearby at Swadlingcote, where large-scale trials and demonstrations could take place to introduce new equipment to the operational management of the mines. At this time, after the two national miners' strikes, and faced with the escalating charges for oil, the national requirement was for increased output of coal if possible. The efforts for change focussed on three main themes — mechanisation of coal face operations, centralisation of colliery layouts and remote control and monitoring of equipment and the environment.

In the Doncaster Area, Jack Wood pursued these objectives with capital schemes, as all the mines had significant reserves and were long-life. He used every opportunity to publicise this policy to the workmen and their union representatives. He expected that they would appreciate the financial investment being made and that the payoff would be better industrial relations. Steam winders were replaced by the latest electrical winders; new coal preparation plants were built, and other coal preparation plants upgraded; central conveying schemes were installed. On the new coal faces, heavy duty shearers and powered supports were installed. It was on these new coal faces that the frustration for management was displayed. Initially we would see achievement of the potential of the new equipment, but then the performance tap would be turned off and the output levels would be reduced. It was a great disappointment to Jack Wood that the investment of money into the pits impressed the manpower very little and that there was no general upsurge in performance to match the investment.

Each year, a day-long seminar was held for the Area to discuss the technical developments and financial performances. This was attended by representatives of the NACODS and NUM union officials, along with colliery management from each pit. It was usually held at the Doncaster Racecourse, which had excellent facilities for such events. The day was well organised, with presentations from different senior members in the Area team. It was chaired by Jack Wood, and throughout he tried to encourage those attending to make contributions to the discussions which took place in a positive way. Despite his efforts there were often some inputs that focussed on the themes of militant trade unionists railing against the Coal Board policy and its management. At one of the later seminars that I attended, I told Jack Wood that I would end my contribution with a challenging finale. After explaining the latest technology that we were installing in the mines, which was very impressive, I ended my piece as follows:

'My grandfather was a miner in the early part of this century. He talked to me when I was a lad, and described his life in the pits. I have always been sad that he died before I could take him underground and show him how we mine the coal today. I would have been proud to show him today's mining systems, but I would be ashamed to have to tell him how poor our performance is when using them.'

Mine was the last contribution before the lunch break. I was told by some of my Area colleagues after the lunch break that some of the union men would have a go at me during the afternoon session. I have no recollection of what they said, so it must have been relatively mild. The truth hurts when you are a trade union militant seeking to destroy an organisation.

Bretby was making strides to develop new control systems for the industry. When they developed and tested a pilaster that would measure very small changes in the percentage of methane in an air stream, it was not long before those measurements could be recorded in digital form. Here was the technology to monitor a mine environment such that accumulations of methane could be indicated to a central control room and any unusual changes passed on to management for action.

In a similar development, conveyors could be monitored and controlled remotely from a surface control room. The technology was complex and involved multiple safety features on the conveyor drives. But there was also technology on the fixed conveying equipment, like hoppers and chutes, to make sure the coal was flowing freely. Much later, cameras were fixed at transfer points so that the control rooms on the surface of the mine could see how the coal was flowing.

The control systems for environmental monitoring and central conveying were designed and specified by specialist engineers at Bretby and were named Minos Control Systems. The software was written at Bretby, and the trials of any software upgrade in the systems at the pits, was managed by them. Maintaining one system in the Coal Board for this technology had great advantages, but there were many companies who tried to get into the pits with their own designs.

The Coal Board tackled this problem by having a National Committee which oversaw remote monitoring and technology industry-wide. It was a significant body, which had representatives from many interested departments of the industry. Jack Wood chaired this committee and he co-opted me onto it. As with all major developments of new technology in a big industry, there are many companies and organisations that wish to be involved and they often approach the operators in the industry and offer technology that they claim is much better than the existing applications. It might be marginally better, but it always requires costly support to maintain any new application. With the Minos systems used throughout the coal industry, any technical upgrade could be quickly applied through all the pits. Jack Wood was a valiant Chairman and insisted that the Coal Board operate its standard system. He even left the meeting on one or two occasions threatening to resign when members were pushing for a change of policy. It was well justified. As environmental monitoring was widely spread throughout the industry, it was rewarded with no further incidents of methane explosions for the remaining 40 years of the life of the coal industry.

There were several technical issues which were specific to the Doncaster Area. The surface land around Doncaster is only a few feet above sea level. How was it possible to lower the surface when extracting the coal without increasing the risk of flooding? Before the coal faces started production, the surface was prepared for them. Any water courses, rivers and canals had their banks built up and strengthened. The farmers' land was then drained down to the nearest water course and an Archimedes pump installed to lift the water up into the drainage system. This procedure was applied for some mine workings to allow more than one coal seam to be worked and it resulted in areas of land around Doncaster being very near sea level. Under normal circumstances this is quite satisfactory, but in a major flood it can require additional pumps to be installed to remove some collections of flood water. Planning and commissioning of this work was carried out systematically without any problems with the local authority or the land owners.

In one case it was necessary to stop working a coal face to avoid subsidence. The coal face from Rossington colliery was approaching the Finningley airport. The airport had the longest runway in the UK and was used by the RAF at the time. In a meeting, I questioned the need for this coal face to be stopped as the surface would only be lowered a few inches. It was explained to me that it might only be a few inches, but if it prevented the very large hangar doors opening, the V-bomber squadron, a major element of the country's defence programme, would be grounded!

Another feature of operations in the Doncaster Area was the hazard of spontaneous combustion at pits working the Barnsley seam. To avoid it arising, roadways were sealed off to stop air leaking from intake roadways through to return roadways. Spontaneous combustion starts when coal naturally heats up if it is fed by some air. The airflows in these mines are monitored continually, and very small changes in the carbon monoxide gas percentage is the indicator that shows when something is going wrong. There are experts at the mines, and in the area, at tackling these problems.

The method of solving a heating problem is to identify where it is, inject cement into the area to kill the heating and then monitor the air samples. You know you are successful when the samples return to normal.

I recall one heating incident at Markham colliery which was potentially very dangerous as it was near the pit bottom, and if it escalated into an open fire it could close the pit. Ted Robson was on holiday, and so I was in charge of operations. We had the best Area experts on the job and the colliery manager was very experienced with spontaneous combustion incidents. The heating had been tackled for several days without success. I visited the site and could not fault what was being done, but the injections were not showing any progress. In frustration, the colliery manager drilled a hole at the opposite side of the roadway to the one where all the action was. He hit the heating with the first hole and it was quickly put out. Why had all the attention been on the wrong side of the roadway? It was found that the gases from the heating had filtered in the roof across to the opposite side of the roadway where they were being measured. Strange things can happen with airflows underground at times. There is an immense sense of relief when a solution is found after several days of fighting such a heating problem.

When I was in Doncaster Area, I had my first of a series of excellent PAs. She was an Irish lady called Patsy Daniels, whose husband was a professional comedian. Patsy collected any funny stories that were circulating and which could be incorporated into her husband's act. She was amazingly fast at shorthand and typing; she could take away notes on a long and complicated technical meeting and return with them all typed up ready for distribution a couple of hours later.

Ronnie Butcher was the manager at Rossington mine. He operated with his consistent style of a rough, tough man who feared no one, and a vocabulary of expletives rolled off his tongue continually in conversations and in meetings. Early in his career Ronnie Butcher was mentored by Charlie Turner and I worked for Ronnie on odd shifts at North Gawber Colliery in my early career.

Ronnie's secretary, whose name I forget, was a very smart lady in her early-thirties, who looked as though she had been through an international finishing school and she spoke with a BBC accent. Once, when in a meeting in the manager's office, she was sitting next to Ronnie and taking notes. It was a spiky meeting, with exchanges between the union men and the manager in colourful, descriptive language more appropriate to a coal face than an office. The secretary calmly sat there smiling and taking notes. I remember wondering how some of those expletives were expressed in shorthand! Thinking about it now it would have made an interesting TV programme!

At the end of my spell in the Doncaster Area, I reflected on what impact the experience had had on me, and how I would describe it. Speaking at a social event back in the South Yorkshire Area I described it as 'Doncastration', a life-changing experience which affects one's body and one's mind, and recovery from which is a slow process! It got a good laugh.

# Chapter 15

*South Yorkshire Area, Deputy Director Mining*
*1977-1981*

Interviews for the post of Deputy Director Mining were held at Hobart House, in London, by a panel chaired by Norman Siddall, Deputy Chairman, supported by John Mills, another Board Member, along with the Head of Staff Department and George Hayes, the Area Director in South Yorkshire Area. George had been the successful General Manager of Kellingley Colliery — a new mine in North Yorkshire Area. He had been promoted as a Production Manager for a short period in North Yorkshire Area before being moved as Area Director in South Yorkshire, to replace Eddie Hoyle who had retired.

I have no recollection of any special features of the interview, but I was known to all the panel except George Hayes, as I had never worked in the North Yorkshire Area. My main competitor for the appointment was Ken Moses, who was at Headquarters in the mining department. There was a rumour that he had got the job, but I got the all-important blue envelope.

In many ways, returning to the South Yorkshire Area was really like going back home. I knew all the senior staff in the Area, and how it functioned. I knew a lot about the pits from talking with the other Production Managers. One challenge for me would be sitting on the opposite side of the table at the meetings from these Production Managers.

## A home of our own

This appointment also led to a major change in our family life. Enid and I went out to dinner with George Hayes and his wife, at a hotel on the outskirts of Pontefract. It was a useful evening, as it helped me to get better acquainted with George away from the workplace environment. Later in the meal, George raised the problem of my Coal Board house in Hemsworth. With three teenagers, we were short of space and I had been negotiating for a few years with the Estate department to have an additional room constructed over the garage. The trouble was that I kept moving between the different Areas in Yorkshire as I gained promotions, and the house took quite a time to follow me.

'I'll have to try and sort out your house problem. It might not be straightforward, though, as you are now on Headquarters staff, not Area staff,' said George.

'We are so fed up with the delays on this that we are tempted to buy our own house,' said Enid.

'That was the best thing we ever did,' said George, supported by his wife. They had moved into a large house in Kirk Smeaton, which was sited such that he could work in any part of the Yorkshire Coalfield.

When we got home that night, Enid opened the local paper. Vissitt Manor was advertised for sale. It was an old, Grade 2 listed manor house, parts of which went back into the 1500s when it was left in the will of Archbishop Holgate of York. It was near where we lived, but it was down a farm lane, which had a slight bend in it, so we could not see Vissitt Manor from the road. Enid had always fancied a house with some character, so she went to view it. Shortly afterwards I got a call to come home as soon as possible. There was plenty of history, but there was also a new AGA in the refurbished kitchen, something Enid had always had on her wish list. It had been a farmhouse for centuries, so there were lots of associated buildings; it even had a bell tower to call the men in from the fields at meal times and at the end of the day. There was an acre of land, with a large cellar under the house and a massive deep well in the middle of the garden which held many thousands of gallons of water drained from the house. It did have major challenges, so we had a family meeting to ensure we had a team effort for the restoration plan. The family decided to go for it, and it took us only seven weeks to complete the transaction, due to support from the Coal Board with an assessment of the buildings, and loans to finance the purchase.

## Vissitt Manor, the structure, 1977-1982

A listed building, parts of it going back to the 1500s was bound to have some structural weaknesses. It had two stone chimney stacks. Both had to be re-pointed and strengthened with access to them by elaborate scaffolding. We had the help of Enid's brother and his son, Robin (my Godson to help with this.) All the ridge tiles had to be re-pointed, that was my job during a few days holiday. The boundary wall to the main garden was a single line of barb wire. It was not very effective, and the farmers shire horses in the adjacent field, paid us a few visits through the barb wire. When we dug down we found the base of the original brick wall, and all the bricks were buried under the sub-soil. That was resolved over several years with brick cleaning (done by the Grandmas!) and brick laying done mainly by me, with other helpers, over several holidays.

A major stone wall, eight foot high, divided the garden into two parts. It had to be built up to its finished height and capped. The stone was obtained from a stock acquired by my father before he died and fetched to Vissitt in his pig trailer. Enid did the difficult task of sorting out the stones to match them and form the bond between the two courses.

Within a few months of us moving into the Manor House an architecture student asked if he could visit the house to see, and photograph, the timbers supporting the roof. As well as the original timbers, which were impressive, fourteen additional trusses had been added to improve the strength of the support system.

He did point out to us that one of the main timbers was fractured and required attention.

This needed some expertise applicable to old buildings. I located two civil engineers, one of whom lived in a very old house, and got them to visit the Manor house and see the problem. They recommended that a steel frame was built, secured into the walls across the roof area, which were 19 feet apart. From the frame the fractured beam could be jacked up to its original position and supported with additional timbers. I asked them to draw a plan of the steel frame and arrange for its manufacture. To allow for handling the girders up into the roof area the long beams were cut in two and had to be connected with substantial connecting plates when there were in position.

I realised that this would be a tricky job needing strength and skill, so the call went out for all the kids to be available and Enid's brother and his son to provide jacks and drills and lifting gear. A Sunday was allocated to carry out the work. On the Saturday night we managed to thread all the steel girders through the roof onto the site. Brynnen also did a bit of speculative drilling of four holes into the walls for the main girders. He tried to measure the position of his drilling to fit the final position of the girders.

On the Sunday morning the build-up took place without any hitches. Bryn's holes were in exactly the right place and the steel beams slotted in with no problems. By lunch time the repair work was complete, and the long beams were cemented into the walls. A job well planned and well executed! There have been no further problems with the roof over the last 40 years.

### Vissitt Manor, the garden, 1977-1980

The garden was covered with thick grass and wick grass. While the grass was cut down with a rotary mower, to recover the soil to make a garden was very difficult. Our objective was to reclaim one square yard per day.

The area beyond the wall was converted fairly quickly into a hen run and an area for the bees. A shed for the hens was transferred from my Dad's and built, again with help from Enid's brother and Robin. A fence had to be built to contain the hens and keep them away from the bee hives. This fence became an important part of the soft fruit production. Along it was set several blackberry plants (the blackberries were huge and were perfect for jams and jellies.) Also a length of the fence was set with Sunberries (a cross between a raspberry and a blackberry) Again a wonderful fruit for jams and jellies.

## *South Yorkshire Area, office organisation*

The domestic arrangements for my office at Wath-upon-Dearne were very good. Alongside the office was a large bathroom with excellent shower facilities. This meant that I was able to visit local mines in my pit clothes and come back to my office to shower and change. If I was very dirty and my boots were also caked with mining dirt, I would take them off and leave them with my driver to sort out. Harold Owen was a wonderful driver, whose aim was to make my busy life as easy as possible. Walking up the stairs to my office in my stockinged feet didn't embarrass me; the staff should realise that our business was underground, and it was tough and dirty. I tried to get three visits underground each week, on the premise that 'one picture was worth a thousand words'. My production managers knew that I was keen to visit any difficult situation at their mines. Support and advice to the on-the-job colliery management is better than just hoping a problem will go away.

I was also blessed with another excellent PA, called Audrey Harper, who had been PA to a National Board member. Her forte was the brief she prepared for any conference or major meeting. The file was meticulously made out and relevant papers included for every item on the agenda.

A central control room in the Area gathered reports from all the mines throughout the day. The young men in that control room were keen to record when mines were doing well and were suitably restrained when the performances were below par. To keep myself up to date I had a call from them each evening giving me the details of each mine's results for the afternoon shift. If I was out during the evening one of my children would write down the figures. This led to me being taken to task when I came home! 'Just look at the performance from X Pit, Dad. What on earth is going on?'

If there were any incidents or accidents, the Control Room knew who to inform and who to call out to support the colliery management.

## The pits and the business

The South Yorkshire Area was a large business, turning over more than £300m per year. It concentrated on producing high-quality coking coal for the steel industry, which was in demand at that time. The saleable proceeds per ton were higher than in other Areas, which were primarily supplying the electricity market. The Coal Board monitored the performance of the Areas with accountability meetings in London each quarter. The team was made up of George Hayes, Area Director; myself, Deputy Director Mining; Martin Shelton, Deputy Director Administration; and Tony Hewitt Area Chief Accountant. We worked well together as a good team. Martin was the wordsmith who drafted all the reports to Headquarters. I used to provide him with the details of the colliery operations, with explanations for any shortfalls and when the pit could be expected to be back to normal. (Later Martin became Secretary of the National Board.) Tony was a contemporary of mine and went to Queen Elizabeth Grammar School a year behind me. He also understood the mining industry and together we formulated the colliery budgets and ensured that they were realistic.

We worked hard in preparation for the London accountability meetings, and tried to anticipate issues that would be raised. As a nationalised industry, the NCB featured in the political arena and from time to time the relevant government department issued instructions to the Board. These included pressures to limit price increases of the saleable coal, to try to limit inflationary factors; and there was always concern at wage increases in the coal industry which could influence other trade unions and their industries. We always provided the meetings with cogent information about any colliery problems that were impacting on the Area results and details about what we were doing to resolve the problems and a forecast of how long the colliery would be affected. George Hayes was very good at farming out contributions to the rest of his team in the meetings, but if he got a question out of the blue he would take it on himself, rather than drop any of his colleagues in the mud.

The nineteen mines of the South Yorkshire Area were spread around the southwest of the Yorkshire coalfield. Brookhouse, in the west, was near Sheffield; Cadeby, in the east, was near Doncaster; Cortonwood, in the north, was near Wath-upon-Dearne; and Manton, in the south, was near Worksop. Three of the mines had an output level of around a million tons a year — Silverwood, Maltby and Manton. The rest were smaller mines: Brookhouse and New Stubbin were working their last reserves, but the management had operational plans to be profitable right to the end, with the full cooperation of excellent workforces.

Elsecar had limited reserves which were only around 50 yards deep. The coal faces were worked under exemptions from the mines inspectorate. Taking out six feet of coal at that depth resulted in significant subsidence such that it was possible to see the land lowering in the fields as the coal faces advanced. The workmen at Elsecar were a breed apart. Some men had worked in their early days when the mine was operated by the Fitzwilliam family and they were aware of the benefits to the community of that family's investments. As young men, they learned to show due respect to the owners and knew how to doff their caps when they met them in the village. To see the teams at work underground was an education. No one appeared to be working very hard, but the results at the end of the shift were way above the performance at other pits doing the same job. There might only be limited reserves left, but Elsecar pit was highly profitable.

It was a time when major spending on capital schemes was taking place in the NCB. There was the first scheduled approval for the Selby Coalfield. Selby was originally planned as a large mine of around two million tons per year, but over a very short time scale that was increased to ten million tons per year. It became the biggest coal mining scheme in the world. To achieve such output levels and move the output from a single site required many aspects of the technology to be stretched significantly. This was achieved only by current mining concepts and equipment being redesigned and uprated.

In the Barnsley Area, major capital schemes were agreed to connect all the mines to just three outlets — at Woolley, Grimethorpe and South Kirby collieries — and treat the coal in large coal preparation plants on those sites.

In Doncaster Area, there was the range of capital schemes as described above (see page 119).

In South Yorkshire Area we were committed to much lower capital spending than the other three Yorkshire Areas. George Hayes and myself met with Jack Godley, the Chief Mining Engineer, to review our projects to see if we should have any additional major capital schemes. George accepted our view that only two major schemes would give a good payoff for the investment. The first one was a scheme to connect Treeton mine to Orgreave mine and treat all the coal in a new coal preparation plant. The second one was to drive a roadway from the surface down to the pit bottom at Shireoaks colliery and use it to convey all the coal to the surface. This new roadway by-passed the shafts, which were of small diameter, and restricted the size of the pit tubs that could be used to wind the coal. There would be the opportunity to lift the total output for the mine and exploit the mechanised coal faces which generally had good working conditions. We were also spending capital on small schemes at all the other mines to improve results.

New coal faces in the Area were mechanised with powered supports. There was a good spread of thicker seams to balance with the thin seams. None of our pits had the very thick seams of the Selby coalfield and mines like Daw Mill in the Midlands.

## *Visitors*

There are always people requesting to visit coal mines. Their reasons might be curiosity or, in some cases, to compare the modern coal industry with the historic mining industry in which their forefathers worked. In the interests of good publicity, we tried to allow as many visits as possible, but they had to be well planned and supervised to ensure the visitors were safe at all times, and that the workmen were not hampered in doing their jobs. Two specific visits during my time in South Yorkshire are worthy of mention.

Firstly, we had to organise a visit by the Foreign Secretary, David Owen. We took him underground at Maltby Colliery, accompanied by the Colliery Manager, George Hayes and myself. At a coal face he saw the shearer was cutting coal and loading it onto the panzer conveyor on the coal face. It is always a relief to see coal being produced as there is nothing worse on a visit than to see a coal face where the equipment is not working for some reason. It was an impressive sight to see a fully loaded face conveyor. We were able to pass through the coal face under the powered supports. It was in a six-foot seam, so there was no crawling on our knees as was necessary in thin seams. David Owen was well received by the men, being a leading figure of the Labour Party Government. He spoke freely to the workmen who were impressed with his friendly and down-to-earth approach. During the visit, David Owen was to be interviewed by Peggy Robinson, a formidable reporter from the Daily Express. She was a smart, impressive, middle-aged lady. When we came out into the pit yard from the shaft, she was there with her photographer. She came towards us and then she stopped, about ten yards away. She had a puzzled look on her face. As we were all side by side in our pit clothes she had no idea which one of us was the Foreign Secretary! David Owen realised her problem and he took off his helmet and gave her a big smile, so that she recognised him. Underground pit clothes are a good disguise. Once I had arranged for Enid to pick me up in the car at Woolley Colliery. I was late out of the pit and I saw she was parked up waiting for me. I went and tapped on the window to apologise and explain that I needed time to get showered and changed. She didn't recognise me and thought she was being approached by a stranger. She got quite cross until I convinced her who I was.

The second visit took place much later, when I was standing in for George Hayes as Area Director. I got a call from Arthur Scargill, asking for a favour. His daughter wished to follow a medical career and she was seeking a place at Sheffield University. Arthur had contacts with senior staff at the Sheffield teaching hospitals. He asked whether I could arrange an underground visit for about 20 consultants. With this number of visitors it would mean two groups going to different locations. When would they arrive at the pit? It had to be on a Friday afternoon when the consultants could arrange to be free! That presented more problems because some pits adjusted the shift times on Friday afternoons so that the men would be home early enough for them and their wives to go to the Miners Welfare Clubs.

A couple of days later Arthur had another request. Could I possibly agree to his daughter joining the visit along with her boyfriend? This was another complication as we would have to arrange for the colliery nurse to accompany his daughter. We never allowed a woman go underground without a female companion.

We settled on Manton Colliery for the visit, and the colliery management team drew up a detailed plan. It specified the time the visitors must arrive at the mine in order to change, in a special section of the pit head baths. Everyone would be provided with a set of underground clothes and footwear, so the pit needed their names and sizes for clothes and boots. The time they would ride the shafts and the time they would board the underground man riders was also specified. The two parties would travel to different coal faces and the visit was timed to arrive back out of the mine before the afternoon shift men had finished their work. Each party had four guides of experienced Manton officials, including the manager and the deputy manager, who would keep them moving and answer any questions. The detailed planning was worthwhile, and both visits were very successful and were on time. I arrived at the mine for when the visitors were back on the surface.

When everyone was showered and dressed they were invited to the conference room for drinks and sandwiches. They showed great appreciation for the visit and said how impressed they had been seeing the coal face equipment in operation and said how much they admired the skill of the miners working on the coal face. A question and answer session followed and as a finale we gave all the consultants a tie in the colours of the South Yorkshire Area. We had puzzled over what to give Arthur's daughter, as a tie wasn't appropriate, so we gave her a miniature oil lamp, which she was very pleased to receive.

A few days later I received a very nice letter from Arthur, thanking me for arranging the visit and acknowledging how successful it had been. He enclosed a wooden NUM plaque for me. Enid displayed it in a prominent place at home. Later, when Arthur was displaying his more militant nature on television, she would turn the plaque round so that it was facing the wall!

When my younger son went to university he thought it would be chic to have an NUM plaque on display in his room. When he graduated and returned home from university the plaque was missing; presumably someone else had decided it would be chic to have an NUM plaque on display.

The visit to Manton by the Sheffield doctors might also have assisted Arthur's daughter, however it was only in 2017 that I learned that she had achieved her objective of following a medical career. She is a GP in Barnsley and I believe that she is highly respected by her patients.

## *The wrong speech?*

The coal industry was a major employer in the mining localities in South Yorkshire and it was to be expected that there would be contacts on many issues between the Area Management and the local authorities. In the South Yorkshire Area, we had especial contacts with Rotherham Council as it was one of our workmen who became mayor for five years in the 1970s. In each case, we made arrangements for the person to have time off from their jobs to carry out their mayoral duties. (I thought that it must have been a shock at the end of their year in office for the man and his wife to resume a normal family life without the large car, with its driver, and the expert staff planning their programme of events.)

One year the new mayor was one of our men who worked at Kiveton Park colliery. He invited me to be the main speaker at his mayor-making ceremony, to which I happily agreed.

Having arrived early at the ceremony, and having a drink at the bar, I looked around and saw not one person that I recognised. I started listening to the conversations, and the speakers were clearly from different industries and occupations far removed from the coal industry. I began to feel a sense of panic. I had got the wrong speech. My thoughts and stories revolved around the coal industry. I spoke to my wife on my fears and she suggested I wait and see when the whole audience was assembled. What should I do? I had no experience of speaking off the cuff to a gathering like this, without any pre-planned notes. I didn't have any options!

We joined the mayoral party and went through to the dining room. It was laid out with a top table and, leading off from it, were three long tables. When we were seated, I asked the mayor who was on the different tables. One was mainly his family and friends, he said; one was chiefly his mates from the pit; and the third was other guests organised by the council. I heaved a sigh of relief and relaxed a little. At least the majority of the audience would know what I was talking about! I don't remember what I said, but I recall it got a few laughs and the mayor seemed happy. He was looking forward to his year in the public eye, full of confidence.

## *Somebody dispensable!*

For about 20 years the payment system in the coal industry had been a day wage, set at different levels for the different jobs. It was considered that it would equalise the payments to the producers on the coal face and the mechanised equipment on the coal face would generate the increases in efficiency. It meant that the only way to increase a person's earnings was to allow overtime. However, over time, it became clear that the installation of mechanised mining systems was not consistently showing increased productivity.

A scheme was devised to link coal face performance to an agreed standard for each particular coal face and any increase above that standard would generate additional payments. There were other factors covered by the proposal. For example, it could be that operations on a coal face might be affected by delays within the mine itself in handling the coal. The bulk of coal to be conveyed to the pit bottom and transported up the shafts to the surface is often measured in thousands of tons per day. These delays were compensated for in the proposals.

Within the Coal Board mine management staff there was support for the proposal, as many of them could remember the days when most operations on coal faces and other work were covered by individual contracts. In those days the men, by their efforts, could affect their earnings and some achieved very high wages.

The NUM was against these proposals. Differential levels of pay made militant action more difficult to organise, so they publicised their objections to the scheme. Arthur Scargill had established good links with Yorkshire Television and his style in interviews made good television. He was able to get appearances on TV whenever it suited him on this major change of NCB policy to introduce incentive pay on the coal face. He seemed to be speaking on the issue every few days and there was no one putting the NCB case.

One afternoon, I went in to see George Hayes and told him of my concerns. 'Arthur has got free rein on this issue, Boss. He's on Yorkshire Television every few days. The Coal Board case isn't getting a look-in. Why can't the Board find somebody who is dispensable and get him on television to put the Board's case?'

George Hayes agreed with me, but he doubted he could do anything about it.

A few days later I got a call to go to his office. What he said to me was a bit of a shock.

'You have to go on television tonight to appear on the news programme with Arthur Scargill.'

My reaction was immediate. 'No way, Boss, you must be joking.'

'I'm not joking. It's Norman Siddall, the Deputy Chairman, who wants you to do it.'

It was a challenge. I travelled to Leeds with Fred Sanderson, who was head of NCB Public Relations in Yorkshire. We planned what I should say in answer to Arthur's arguments. We knew that the slot would only be a few minutes, so I had a few key phrases that I would try to get in. When I went into the Green Room and Arthur saw me, it seemed he was somewhat surprised.

It was a new experience for me and I realised that I had to ignore the cameras and the technology and concentrate on the words. Arthur put his points forward in his usual positive style. I argued in favour of this policy and had answers without being lost for words. Fred thought it had gone okay, and Enid was complimentary. There was certainly an adrenalin rush when it was all over. The exposure risk came later when some people said they had seen me on television. I wasn't sure it was them that I was trying to influence.

It then became serious. It was decided to set up a small team in Yorkshire who would be responsible for appearing on local TV or local radio on any issues affecting the mining industry. It was led by Mike Eaton, with myself, an industrial relations expert, and a person who could cover marketing and administration issues. We were sent on a training course run by a distinguished TV presenter who had appeared on Panorama. He emphasised that any interviewer is desperate not to have a guest who dries up. We learned that there is scope to influence what questions are asked and what subjects are banned. It was essential to keep in mind how long the appearance would be so that you could focus on the key issues. I recall one appearance with Harry Gration on Look North when the slot was reduced drastically in mid-programme. He whispered that we would only have 90 seconds. I whispered back 'questions one, three and five.'

Many issues in the coal industry were put to a ballot of the workforce. Arthur Scargill's approach to ballots was to work hard to get his views over by having mass meetings around the country. Mike Eaton realised that his team had to know what line Arthur was taking and the arguments he was using. In Yorkshire, the meetings were always held in the City Hall in Sheffield. We dispatched a couple of young men from the NCB Public Relations team in Doncaster who had press cards, with instructions to sit on the front row and take notes of what Arthur said. They then had to come back and report to us. It was a late-night meeting with them.

Initially they were amazed.

'He was awesome. He had the audience on their feet and he swept them along. We have never seen anything like it.'

We knew that Arthur had immense skill as an orator, speaking to an audience and inspiring them. But what had he said?

They gave us details of his arguments and his predictions if incentives were reintroduced on the coal faces. We knew that if Arthur was to speak in other coalfields at mass meetings it would be the same speech. (In some cases, a section of the audience would also be the same. Arthur was able to organise a travelling circus that followed him around.)

Also we knew that Arthur would stick to his script. He was not a debater who would react to other ideas. It was the same during negotiations — he wouldn't compromise. Meetings between the NUM and a Coal Board unit would have an agenda of several items, say seven. It would be the objective to agree a compromise between the two parties. Arthur would not compromise and would be prepared to leave the meeting loosing six cases to one rather than changing his position.

Our aim as a PR team was to agree what lines we should take to counter his arguments and challenge his predictions about the application of incentives in the industry. Arthur may have had standing ovations at his meetings, but the men were wise and realised that incentives could affect their pay packets. So, they voted to accept them, and incentives were introduced and were successful. What puzzled HQ in London was that if Arthur lost a ballot they expected him to go away and hide; but he returned to the fray as though he had won. They didn't realise that Arthur was a true Marxist: he was fighting a war to change the world; and losing a battle he just put down to experience. It was winning the war that really mattered.

Appearances on TV and local radio were usually live. Meetings with the press were less predictable for results, but Mike Eaton was very good with reporters. I had one unfortunate experience with Channel 4. I was asked to record an interview on the industry for a programme they were preparing. A very quiet and demure young lady with a soft voice asked me some questions regarding the future technology in the industry. I took this as a green light and I waxed lyrical about new technology and illustrated it with major examples, some from developments in the Selby coalfield. I thought there was good story to tell and that I had put it forward very positively. When the programme was screened the questions to me were re-recorded by a male interviewer with a challenging and accusing voice. The theme of the programme was that new technology in the coal industry was being introduced to destroy jobs. I had been conned, and my colleagues recognised this. Although it's more nerve racking to appear in a live programme, at least no editors can manipulate what you have said.

### An added responsibility

It was usual for the appropriate Coal Board technical staff to analyse major incidents in other industries, in the UK and overseas, and consider the risks of such an event happening in the coal industry.

On 1 June 1974, the Nypro (UK) site at Flixborough was severely damaged by a large explosion. Unfortunately 28 workers were killed and a further 36 were injured. Offsite property was damaged to varying degrees and there were also injuries to 53 other people in the community. It was accepted that if the incident had occurred during the week, when the site was fully operational, rather than on the Saturday afternoon, the casualties would have been much higher.

As with all major incidents, it was a combination of factors which caused the explosion. One of the reactors in the plant was found to have a serious problem and was closed. A bypass pipe range was installed to keep the rest of the plant in operation. During the afternoon of 1 June, a 20-inch bypass range ruptured, allowing a large quantity of cyclohexane to escape and form a massive vapour cloud which ignited and started numerous fires on the site. These burned for several days and hampered the rescue work.

Could this apply on any NCB site? The NCB Mines Rescue Section of British Coal had been involved in the incident investigation. They considered that the Manvers multi-unit plant had constituents that could cause a serious explosion. From the coking plant there were a range of liquid fuels produced and stocked on the site.

After all the investigations and reports, the Head of Mines Rescue, Gordon Jolliffe, had a meeting with me to inform me that if there was a major incident on the Manvers site, I would be the official in charge of handling the incident. I would have to coordinate the rescue efforts and manage the impact on the other mining operations at Manvers and in the locality. The assessment of the gravity of an incident on the Manvers site was that it would be many times more serious than that which had applied at Flixborough. If the wind was from the West, Mexborough would be devastated; if the wind was from the East, Wath-upon-Dearne would be devastated. We spent some time considering what would be needed in response to a Manvers incident.

In retrospect, neither I nor anyone else in the coal industry had to perform in this capacity. I was fortunate to conclude my career in the coal industry without being directly in charge of any major incident involving multiple deaths. Having seen the effect such incidents had on even the toughest managers, I believe I was very, very fortunate.

### *George Hayes' accident*

George Hayes was having some work done on his house. He was concerned about the progress one day, so he went home for lunch so that he could inspect the work. He didn't come back when expected, and then we got information from his PA that he had fallen from the roof of his garage and was injured. The next day, Martin Shelton and I went to see George. He was resting in a darkened room and looked seriously incapacitated. We had no idea what the outcome would be. We told him to forget about the South Yorkshire Area and concentrate on getting better. Martin spoke to Hobart House in London and informed them that George might take some time to recover. He also informed them that we would be able to manage the South Yorkshire Area and cover for George's absence.

There followed a period of over two months until George was able to resume his work. I recall that period as being a haze of activity. All the

Production Managers raised their game and kept the pits supervised and directed; the Area team worked together and took the decisions and reacted to Coal Board policy instructions in the way we considered best, and in line with George Hayes's views, as we recalled them. There were all the special events of the Christmas period, with dinners at which I had to speak; and it was the time of year when all staff had their 'blue letters' advising them of their increase in salary for the next year. There was also a small number of staff retiring at the end of the year, who had to be thanked for their services to the industry and given our good wishes for their retirement.

It was a slow recovery process for George. He was a very brave man to return to his post. Many men would have retired and called it a day, but he was determined to get back to work. He continued, when he was back at work, to have severe headaches from time to time, and on occasions he had to switch off and rest in his office in semi- darkness. We respected and admired him, and gave him our full support in doing our jobs in the Area.

In the late summer of 1981, I was asked to see George in his office. He told me that I had to attend for an interview at Hobart House with the Deputy Chairman and others, in two days' time. I asked the reason for the interview, but George said he hadn't been told, and I believed him. There had been several changes in senior staff in the Yorkshire Coalfield over the past year: Michael Eaton had moved to North Yorkshire as Area Director, Albert Tuke had moved to Doncaster as Area Director, and John Keers had taken over as Area Director in Barnsley Area.

The interview was very short. Bill Forrest, who had been in charge of the Selby coalfield, had retired much earlier in the year and had not been replaced. He had piloted the project from its concept phase through its detailed public enquiry period with local authorities and into the development phase.

The proposal was that I should take over as Deputy Director Mining (Selby) under Mike Eaton who was Director of North Yorkshire Area. Dennis Bousted was Deputy Director Mining for the other pits in North Yorkshire. On the same date, 1 October 1981, Peter Hall was to start as Deputy Director Administration in the North Yorkshire Area. I had worked with Peter Hall, who was staff manager in Doncaster Area. There was little debate at the interview, I just said 'yes'. To be entrusted with responsibility for the biggest mining project in the world at that time seemed a vote of confidence, even if in title alone it was a sideways move.

The following morning a call came through on the telephone in my South Yorkshire office. It was Mike Eaton, informing me that he had wanted this appointment for some months, but getting me released from South Yorkshire had not been easy.

Before I left South Yorkshire, I spent a day in North Yorkshire and saw the offices that had been prepared for Peter Hall and myself. I also conducted interviews for my PA. Again, I scored a winner, as Sandra Steele was a lady who had presence and authority and was able to deal with hosts of distinguished visitors from archbishops to senior politicians and company VIPs. She looked after the office when I was out on the sites and knew how to plan my diary to squeeze the maximum output from my time.

# Chapter 16

*Deputy Director Mining, Selby*
*1981-1985*

*Background*

Bill Forest had piloted the Selby project from its concept phase, through its detailed public enquiry period with local authorities, and into the initial construction phase. Bill was ideal for the early, patient deliberations and conception of the Selby story. He was a man of presence, and with his rich speaking voice and his white hair, any audience would assume his remarks were made with a large dose of wisdom. He knew that solutions to operational and environmental constraints would only be found if the boundaries of technology in mining were pushed up from their perceived standards in the mid-1970s.

Bill had assembled consultants and architects who were capable of achieving the objectives, but it was inevitable that some perfect solutions would interact adversely with the solutions proposed by other specialist engineers. All the expertise had to remain focussed on achieving a technically sound, but prestigious, solution to the problems. During the four years I was in charge of the project, often the emphasis of my management role was to recall engineers from the extreme position they were taking on particular problems and their solutions. They needed to introduce the stark reality of time and cost into the decisions they made if they were to reach a sound compromise which defined the right technical solutions to be applied in the coalfield.

The NCB management team at Selby wanted to get through the development stage as quickly as possible and see the coalfield into production and generating a revenue return for the high level of capital investment. I was fully behind that objective and was happy to focus my efforts in that direction. While there were to be many new issues to affect my role at Selby, I was quietly confident that my experience in the coal industry over the years had been sufficiently varied to prepare me for the task.

## *The organisation*

The organisation to manage the project consisted of a main consultancy firm, W S Atkins, whose leader was Peter Hammond. There were two different architects: one covering the Gascoigne Wood site, and the other the five shaft sites. In a similar way, there were two different quantity surveyors monitoring the spending on the different sites: Wakeman Trower, of Leeds, covered the Gascoigne Wood site; Faithful and Gould, also of Leeds, covered the five shaft sites. I had a team of Coal Board support staff in the various disciplines to inspect and monitor the operations on the sites. Regular meetings were held to chart the progress of the works and also to evaluate the spending in relation to projections by my planners. The project had an annual financial allocation from the NCB and the cumulative spending and progress with works was reviewed in the quarterly accountability meetings by the Board with the North Yorkshire Area management. It was accepted in the Coal Board that the Selby Project was a large and complex technical challenge but, by the time I arrived, there was confidence that the organisation and technical experts available were appropriate to achieve the objectives.

## The concept

Planning permission was granted for the Selby coalfield in April 1976. It covered 110 square miles in the Vale of York. The coal reserves had been proved by 60 boreholes and 300 miles of seismic surveys, and established total reserves of 2,000 million tons of coal, with 600 million tons in the Barnsley seam. The consent for mining the coal applied just to the Barnsley seam.

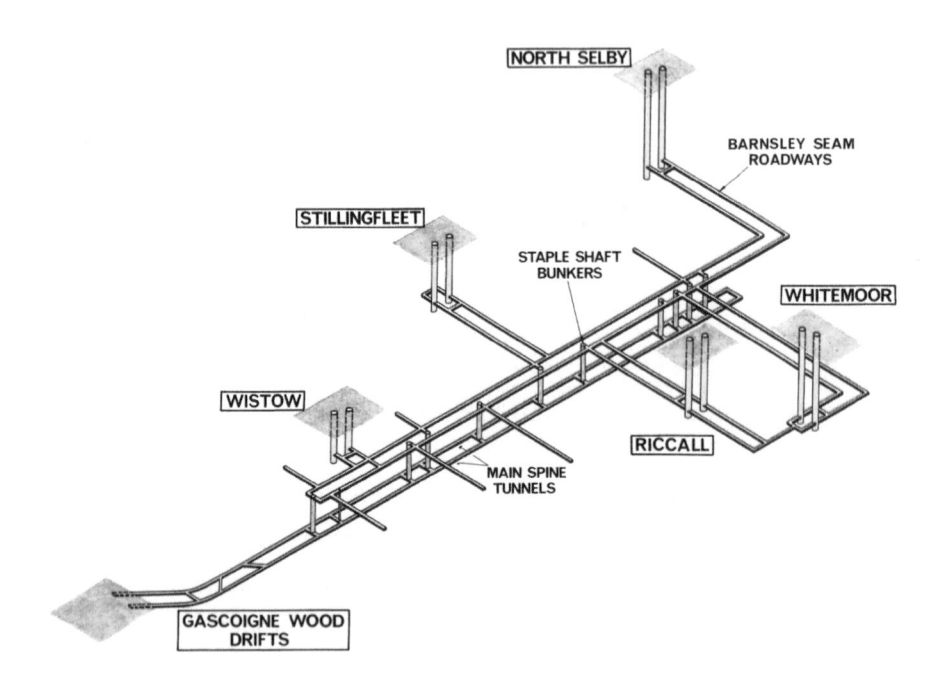

*Schematic layout of Selby coalfield showing shaft sites and Gascoigne Wood disposal point.*

A fundamental decision was that all the coal should be brought to the surface at one point and dispatched by rail to the power stations. That surface site was at Gascoigne Wood, a disused marshalling yard, just outside the mining area but conveniently alongside the main rail line route to the power stations. In the coalfield area, the Barnsley seam was at a depth of 250 metres in the west to 1,100 metres in the north east.

Two access drifts from the surface at Gascoigne Wood were driven at a gradient of 1 in 4 for 800 metres to a position below the Barnsley seam. These were driven through water-bearing strata and systematically the strata was frozen solid and the drifts were driven forward and supported with circular arches and cast iron 'tubbing'. The strata behind the tubbing was injected with cement to seal the water back from the finished drifts. The sequences of extraction and support were successful and both drifts were watertight, with just occasionally a 'pin prick' leak that was easily sealed.

The two drifts were then to be extended as spine roads at a gradient of around 1 in 20 to a final length of 14.8 kilometres from the surface.

A second major decision was that there should be a single conveyor in each of the spine roads to bring all the coal to the surface. This reduced the need for tandem conveyors underground and the enlarged excavations to house them; it also simplified the electrical distribution underground and all the splicing equipment and installation of the conveyor belting would be done on the surface. To bring all the coal to the surface along the spine roads required conveyors of exceptional size and horsepower.

The concept became a reality when it was agreed to power both conveyors with the NCB-rationalised winder motor designs using thyristor controlled, variable speed direct current machines. The option to use a slow speed start-up procedure was a great feature of these conveyors.

Two different types of conveyor were used:

North Spine conveyor — cable belt; maximum load 2,750 tonnes per hour; maximum speed 27km/hr; belt width 1,050mm.

South Spine conveyor — steel cord belt; maximum load 3,225 tonnes per hour; maximum speed 30km/hr; belt width 1,300mm.

Conveyor Drive Motors:

North Drive — 2x variable speed 4,133kW.

South Drive — 2x variable speed 5,050kW.

To make the system flexible and avoid overloading, variable speed, accelerator conveyors would feed the coal at the required rate onto the two trunk conveyors.

The surface drive houses to hold these massive conveyors had to have very secure foundations. The steel structure of the buildings was supported on piled foundations tied together with a grid of tie beams. Plinths to support the main drive motors, producing 10,100kW had to be designed to withstand a horizontal pull of up to 1,950 tons at a height of 10m above ground level.

An additional electrical feature had to be built. If the conveyors were travelling at full speed and full load and they were switched off, the electrical surge would trip out much of the electrical supplies in Yorkshire. To avoid this problem a bank of switch gear was installed to act as static compensators to absorb the power surge.

As the mine sites would not be handling the coal, their role would be to give access for the men, and the mining materials they needed, to drive the roadways for the pit infrastructure and open out the coal faces. There were height restrictions on the shaft headgears and any surface buildings. There was also a requirement to limit the acreage taken for those sites. Wistow mine was the smallest site, with only 29 acres. The buildings were clad in colours and shapes so as not to look like mine sites and the landscapes were interlaced with gardens and shrubs. The material from the shaft sinking was formed into bunds around each site which acted as a partial screen for the buildings. One visitor I was escorting around the completed Wistow surface buildings asked where the trolleys were kept — he thought it looked like a posh supermarket. That was confirmation to me that the promises made to the community of ensuring the mine sites did not blight the landscape had been achieved.

Much later, when the Wistow mine was fully operational, there was consideration of what to do with the grassed bunds. Some very expensive suggestions were made by landscape experts, but it was decided to return the land to its original use. The local farmer had sheep on the bunds and they provided a pastoral aspect to the site.

## The Selby Project when I arrived on 1 October 1981

At Wistow, the shafts, both with a finished internal diameter of 24 feet, were completed to the Barnsley seam in September 1981. Work was in progress to install the shaft furnishings at the surface and underground, to be followed by building the winding engines. The upcast shaft headgear was slip formed over one weekend in reinforced concrete, which gave a welcome impetus to the surface civil engineering work schedule. The downcast shaft had a 16.5 tonne/170 men capacity, single cage and counterweight-balanced winding system, with the tower-mounted, multi-rope friction winder located directly over the shaft. To carry out this design, which complied with the height restrictions on the site, the motor rotor was carried on an extension of the drum shaft with the motor stator supported on secondary beams. The deflection between the motor rotor and motor stator had to be limited to less than 0.3mm. To achieve this standard, the winder-supporting plate girders were 1.6 metres deep, with 75mm thick flanges, and weighed 18 tonnes each. This is an example of a refined engineering design being required to meet the planning consents. The downcast shaft was excellent, and allowed the transport through the shaft of built up machines which could be moved under their own power to their point of use underground. In a normal mine, to strip down a large machine on the surface and rebuild it underground ready for use, could take several days using the craft teams of the mine.

At Stillingfleet mine, the shaft sinking was through the frozen area, with water-bearing strata into normal sinking operations in both shafts. At Riccall mine, the shaft collar was complete, and sinking was down to the frozen strata through the water-bearing rocks. At Whitemoor and North Selby mines, the shaft sinking operations were under way.

At Gascoigne Wood there was activity all around the site. My first site visit was to Gascoigne Wood and I was not impressed by signs of activity. There seemed to be few men around, so I raised this with the senior people with me.

'Just wait until the hooter sounds for the mid-morning break.'

True enough, half an hour later, the reaction to the hooter was that dozens of men appeared at different places around the whole site on the surface. The focus at that time was on the sub-surface foundations. Those foundations were complex and had to be to the required standard to carry the varying loads that would apply when the coal was flowing. It would be months before structures started to be erected in the form of surface buildings.

Underground at Gascoigne Wood, from the surface drift the North Spine Road was being driven as a standard arched roadway using the Titan roadway heading machine. The best performance of 74.7 metres in one week was achieved in October 1981, just after my arrival at Selby.

*The Titan Drivage Machine*

The South Drift was completed, and a massive chamber was excavated to allow the build-up of a Robbins Miner full-face tunnelling machine. The machine was built in Seattle and shipped to the UK as a giant construction kit. The chamber was 40 metres long and 8.5 metres high to give sufficient space to erect the tunnelling machine. The circular cutting head of the Robbins miner revolves with 42 roller cutters to give a circular roadway of 5.8 metres diameter. The pantechnicon behind the miner, incorporating a conveyor to dispose of the dirt and special handling equipment to move the steel support sections forward to the front of the machine, was as long as a British Rail high speed train. It also contained special laser equipment to guide the machine in the required direction and at the precise horizon of the South Spine road. The total weight of the machine and the pantechnicon was 240 tonnes.

*The Robbins Miner*

*Community consultation*

It had been established before I arrived in post that there should be an open meeting, in Selby town, every three months, to report on the progress of the project. I welcomed this, and it was an extension of my PR role in the

Yorkshire Coalfield. The challenge was to report on progress on the surface sites and the underground operations. We always tried to illustrate what was happening with slides. It was to be expected that the people of this rural area would have their own views of what mines looked like and have reservations of what the impact of the coalfield would be on their lives. There were developments taking place to house the miners from West Yorkshire who would transfer to Selby to work the mines when the coalfield was at full production. These plans involved building many houses, which would have a significant impact on the communities where they were sited. There was a mixed reaction to these proposals: there was approval of the population in the area increasing, but apprehension that the nature of the communities might be dominated by the incoming miners. We were able to show slides of the plans for these houses at the community meetings to emphasise that the developments were of high-class houses which would be a long-term asset to the communities. During the development phase the civil engineering contractors and the shaft sinking contractors primarily used their own manpower and staff on their contracts. Those men, in many cases, were itinerant workers, of various nationalities, who moved around to follow the major projects. It is interesting to note, in retrospect, that not all the Yorkshire mineworkers who moved to Selby settled permanently in the new housing provided; some moved back to the West Yorkshire communities where they had been brought up and they preferred to travel to work.

The Gascoigne Wood site was isolated from the communities in the Vale of York, so the large buildings on that site were not a major issue of contention. The mine sites and the impact that they had on the local roads and facilities were more important. When the Wistow site buildings were completed, they generally had full approval from the community.

The vastly important worry was the subsidence that would arise from extracting the coal. The planning consent required the subsidence to be limited to one metre. All the designs for the coal faces at Wistow, in the shallow areas of the coalfield, were based on partial extraction, to limit any subsidence to the one metre. It was natural for me to be questioned on this subject. I could explain all the experience that was available to calculate how much the surface would lower in any particular circumstances if a seam of a known thickness was mined. Selby had been prone to flooding from time to time, so it was possible for some person, probably a dear old lady, to ask the question, 'If it floods again will the water be one metre deeper?' There is only one answer to that question, but it was not appropriate to give it. There actually was a positive answer, which subsequently evolved, and will be described later.

## *Wistow House*

The land for the Wistow mine was bought from a farmer. Adjacent to the site was the farmhouse. It was decided to adapt the house so that it could be used to entertain VIP visitors. A cordon bleu female chef could prepare excellent food in the kitchen, and there was total privacy for any discussions that took place with the visitors.

As the project progressed there was a natural increase in the number of VIP visits, including some international mining engineers and politicians. Some were challenging in their questions about the technology and the costs involved, and I enjoyed these; others were fascinated by the technology, but they didn't relate to the aims of the project to produce coal at much higher levels of efficiency; yet others were just bored! I recall one afternoon visit by a large party from the European Economic Community. Their morning visit had been followed by a large lunch, with copious quantities of alcohol. We were waiting to entertain them at Gascoigne Wood, and we had a series of telephone calls delaying their time of arrival. They eventually arrived over two hours late, some in a serious state of inebriation, and requiring a nap, not a lecture about the largest mining project in the world at that time. I gave them a very short, snippet view, of the Selby project and got them back on their schedule.

## Flooding, 1982

In January 1982 there was serious flooding affecting the coalfield area. I went by car to inspect the scene and it was just a sea of water, with the streams and rivers indistinguishable. It was a terrifying site and brought into focus the fears of the local people of the Selby area about the risk of floods. We discussed this among our Selby team. Was there anything that could be done? We decided to call in the experts and appointed Binnie and Partners, who were river and drainage professionals, to investigate and prepare a report, but our hopes were not high.

However, when we got their findings, we were surprised to see that there was indeed a possible solution. It was proposed to construct a bund on the west side of the river, which would give protection for all the properties in Selby and the adjacent communities. The height of the bund was such that between the bund and the river there was capacity to hold more water than had been produced in any previous flood.

This was a compelling proposition, offering a major insurance guarantee for the residents of the Selby locality. It was agreed that we should undertake this, and finance it, as a part of the Selby project. We had discussions with the local authorities and got their support. It made a good PR subject for the next quarterly meeting with the community. One would have thought that this proposal would have been received with acclamations of praise and goodwill. No way! The people along whose land the bund was to be built were anything but supportive and were persuaded only very reluctantly to allow access. The bund was built and, as a result, the fears of any surface subsidence and increased flooding in the coalfield, were significantly reduced.

## A new railway line

The East Coast main line between Doncaster and the North East, via Selby, ran through the western area of the coalfield. To leave a pillar of coal to support that line would have neutralised a large proportion of the coal on the western side of the coalfield. These reserves were the first coal to be mined at Wistow mine and they were the thickest sections of the Barnsley seam in the Selby project.

It was agreed to build a new line, clear of the western limits of the coalfield, and it would be financed within the project. There had been little experience in British Rail at that time of building new lines. They assembled a team to manage the project and its total cost was £78 million. We had routine meetings with the BR team to monitor the progress and the costs. There were no major technical challenges but, with the flat landscape in the Vale of York, the scheme involved significant road modifications to raise the roads onto bridges over the new line. There were also new bridges over the Selby canal and the river Wharfe.

On completion of the project there was the issue of the land which formed the old railway line. Having bought the land for the new line, I argued that the land should be transferred over to British Coal. It was a sound argument, so the British Coal Estates department took over management of that land. It was used for a cycleway and also some was transferred to the Highways Authority to be used to form a by-pass for Riccall. There were mumblings that the new by-pass would benefit if there was a pillar left in the coal workings to support it. When the historical facts regarding the new railway line and its cost were explained, the mumblings faded away! I had the pleasure of travelling on the first test train from York to run down the new line, and it was a good ride. Now thousands of people each day pass along the line at high speed, with a fine view of the Vale of York, but there are no signs of the coal mines!

## Managing the finances

The routine meetings by the British Coal team and the consultants had to focus on the costs in relation to the progress of the works. In any large project with a budget spend of £3.7 million per week, there are bound to be some people who consider that the project effectively has a blank cheque book. One trivial case was identified when we learned that one of our quantity surveyors held the team monthly meeting with a contractor's quantity surveyors over a lunchtime engagement in a local hostelry; they each paid for the meetings in their turn. That practice was quickly stamped out.

A major issue on the finances arose for the year 1982/1983. The allocation for Selby was £178 million for the year. Both parties of quantity surveyors presented figures that indicated clearly that the £178 million was not enough. I would have to go back to our Board for an additional allocation. I was aware that the NCB had no additional capital fund and that they would have to approach the Government for a supplementary allocation. Such a request would no doubt precipitate an investigation into the project and the way it was being managed. Hosts of people involved in the project would be diverted into answering questions and explaining in detail the build-up of the costs rather than focussing their efforts on getting more work done for the money we were spending.

My chief planner, Jack Stanton, was a man very experienced in project work and the management of costs. I sat down with him to discuss this issue.

'What are we going to do about the annual allocation Jack? There are some strong views being expressed that it is just not enough.'

Jack thought for a few minutes before he answered.

'Well, Boss, in my experience quantity surveyors always like to have a bit up their sleeve. I know the £178 million is a tight allocation, but I think if we manage it carefully, we can keep to that figure for the year.'

'If you think that, Jack, we will work to that figure and we will let everyone know that we have to achieve that result.'

Jack was right, and we did keep the total costs for the year within the £178 million allocation.

There was another case, much later in the project, which required sorting. Major contracts for the shaft sites covered the shaft furnishings and the surface and underground equipment at the shaft side and in the pit bottom. The first two sites, at Wistow and Stillingfleet, were done by Qualter Hall, a local Barnsley company. While Qualter Hall had done a very good job, it was considered that a different company should be used for the third contract at Riccall and it was let to a major national civil and engineering company. They had a young site manager whose objective seemed to be to generate as much money as possible from the contract. At the slightest opportunity he submitted a variation order claiming additional payments. These were contested at the local level, but our QS was concerned at the build-up of the total amount of these claims which had not been resolved. There was no sign that the company site manager would compromise on any of his claims. I was approached by our QS to get directly involved with this problem. This contract was the worst one on the Selby project to have so many individual claims that were unresolved.

I was always told that in any negotiations you should not play your ace if a lower card would win the hand. However in this case I thought I had to play my ace.

I rang a senior director in the company and explained who I was and what was happening with their contract. He was aware of the contract, but not the latest details. I asked him if his company would want to undertake other contracts in the coal industry? Of course, he said that they would. I then explained that my only option now was to write to the Director General of the Supply and Contracts Department for the NCB and describe what was happening on the Riccall contract. I explained that I was confident that he would conclude that the company should not be put on any more tender lists for jobs in the coal industry. This was clearly a shock to him.

He asked me to give him time to review the situation. I was seeking a compromise but, after his investigations, he wiped the slate clean of all the variation orders. On a project like Selby, he could not afford for his company to be tainted by the way their contract was being managed.

There was another case which worked the other way. Anderson Strathclyde was a major national company who did a lot of business providing coal face equipment for NCB mines and also overseas mines. They won the contract to design and build the steel cord conveyor and drive units for the Selby South Spine roadways. It was a complex contract which was stretching conveying technology to new heights significantly beyond anything available around the world. When the design was finalised for the drive units, the conveyor belting and the conveyor structure, it was clear that the costs would be above the tendered prices. There was no alternative available for other equipment, so we had to agree a new price, otherwise the company would have gone out of business. Mike Eaton and I, with support from the Supply and Contracts Department, had meetings with senior management of Anderson Strathclyde and a new price was agreed. It effectively became a cost-plus contract, where all the relevant costs were charged, and the Company was paid a small mark-up. The detail of this arrangement was carefully monitored by our QS team. The steel cord conveyor was a technical success and contributed to some very high outputs from the coalfield.

### Initial installation of trunk conveyors, 1983

The progress of the tunnelling machines in the spine roads was critical in establishing a connection to Wistow to take coal from the first coal faces.

In 1982 the North tunnel averaged 43.91 metres per week and the South tunnel 90.25 metres per week.

These performances allowed the connections to be made to Wistow from the spine roads and for the trunk conveyors to be installed to that point. The one chosen first was the cable belt, as this was a standard product. The details of the steel cord conveyor and drives was delayed, so it had to be commissioned later. To release the North Spine road and Drift for the installation of the cable belt, a connection was made between the North Spine road and the South Spine road, so all the materials from both spine roads was conveyed to the surface along the South Spine. The connection to the Wistow workings was made via an enlarged borehole which remained the route for the coal for an extended period.

The cable belt conveyor and the steel cord conveyor were installed for the first 5,000 metres of the roadways. Installation of the steel cord conveyor started in October 1982 and the Wistow coal from the first coal face was brought to the surface on that conveyor in June 1983.

## Two technical challenges

Going forward, the Robbins Tunnelling Machine entered a stretch of strata impossible to work with the machine. If anyone had set out to design a piece of geology completely inappropriate for a tunnel boring machine, they could not have done better than the strata at that point in the South Spine road — it was riddled with variable weak bands of shale measures crossed by numerous small faults.

The solution to this problem was to halt the Robbins machine and extract by hand a length of roadway above the level of the roof of the tunnelling horizon. A series of one metre square cubes were mined and supported by a steel frame. The frame was then filled with concrete. It was a slow process and the excavation was advanced 15 metres until the roof conditions were back to near-normal.

The rescue plan was successful, and the Robbins machine advanced under the reinforced steel and concrete roof, and then continued forward on its way.

In the North Selby mine shaft sinking, it was known that water-bearing strata would be met at levels of the Brierley rock 486 to 513 metres, the Ackworth rock 524 to 583 metres, and the Shafton sandstone 594 to 631 metres. The water would likely be at high pressure. The size of these challenges is best illustrated by giving comparisons of water pressure in other circumstances. The water pressure on the sea bed of the North Sea, at a depth of 168 metres, is 238psi (pounds per square inch); the greatest depth a nuclear submarine would operate at is 300m, where the pressure could be 426psi.

I went on a visit to the North Selby shaft sinking when I was told the pre-sinking borehole had tapped the water-bearing strata. On the floor of the shaft the Master Sinker told me that they had struck the water.

'Are you sure?' I asked him.

'Just stand back,' he replied.

He opened the valve on the probe pipe and the jet of water hurtled up the shaft hundreds of yards!

In the Ackworth rock at North Selby, at depths of 550m to 578m, water pressures of 1,000 psi were encountered.

The flow of water had to be sealed by injecting cement under pressure to form a curtain in the strata to allow sinking to progress. Different boreholes and standpipes were tried, but the flow into the base of the shaft was not controlled. Also, the base of the shaft sump had lifted by 880mm. It was decided to construct a 5-metre-thick plug at the base of the shaft, and tie it into the shaft walls. The plug had 48 angled standpipes incorporated into it. The hydrostatic load on the base of the plug was calculated at 56,600 tons. In January 1983, a total of 128,000 gallons of Cemex A2 was injected into the initial 24 holes, and a further 27,000 gallons of Cemex A2 was injected through the 24 infill holes using pressures up to 1,250psi.

In late March 1983, the plug was removed and sinking made through the Ackworth rock with an inflow still of around 120 gallons per minute. Sinking continued and the concrete lining was cast at 1,400mm thickness. After subsequent grouting, the flow into the shaft was down to 5/10 gallons per minute. A major challenge solved!

## Gascoigne Wood surface buildings

When the foundations were complete, the surface buildings began to rise up on the site. The most impressive building was the covered stockyard, designed to hold 43,000 tons of coal which would be blended to the required consistency using stacking and reclaiming machines. The building was 300 metres long by a 65-metres clear span wide and 22 metres in height. It had an open end to allow the stacking and reclaiming machines to operate outside as well.

The main foundations for this building were diaphragm walls cast under bentonite, in the shape of a 'T' 2.5 metres wide with a leg 3.2 metres long and 800 mm thick and founded on bedrock approximately 22 metres deep. These foundations were able to resist the combined reactions at the base of the three-pinned arch of 200 ton vertically and 173 ton horizontally. At the time of its construction, the covered coal stocking area was considered to be the largest single-span industrial building erected anywhere in Europe.

When the coal arrived at the surface at Gascoigne Wood it was carried by conveyors in overhead gantries to out-loading bunkers or to the covered stocking area. Other covered gantries carried the coal to surge and loading bunkers over the mine railway line where a 1,000 tonnes train could be loaded in about 20 minutes. The whole site was designed so that the circulation routes for coal, people and traffic can all move freely without conflicting with each other, and the railway trains can be serviced quite separately.

Progress with the buildings at Gascoigne Wood during 1992 was good and indicated that the surface facilities should be commissioned ready for the production of coal from the first coal face at Wistow.

The size of the covered stocking building enabled the assembly of the equipment for Wistow 1's coal face.

*A shearer on the surface*

It was used to familiarise the workmen and officials who would be deployed there, and they were able to move the shearer along the panzer conveyor and operate the powered supports. The section of the seam required all heavy-duty equipment. The double-ended shearer was 400 horsepower, and the face panzer was 30 inches wide. The photograph shows the shearer disc with its cowl. Over the panzer conveyor is the rotating crusher to reduce the coal size, as all the product was going into the power station market which uses pulverised coal.

The powered supports were Dowty 4x450 ton chocks. The photograph shows the fore-poles which can be extended to give immediate support to the roof when the shearer has cut the top coal as it moves along the coalface.

When the face equipment was set up in the massive covered stockyard it did not appear particularly large, but compared with normal coal face equipment it was an exceptional size. The retreat face in the Barnsley seam sections at Wistow was expected to produce coal at a rate of up to 500 tons per hour.

*Dowty powered supports*

## President of Midland Institute of Mining Engineers

Having been an active member of the Midland Institute of Mining Engineers for many years, and having also served on its Council, I was installed as its President for the year 1981/1982. I succeeded Mike Eaton, my Boss in the North Yorkshire Area, in that capacity. The President's role is to chair the Council meetings and the monthly meetings when technical papers are presented. The social events in the year included a dinner/dance, to which partners were invited and was always a sell-out. There was also an annual dinner when distinguished senior persons in the mining industry were in attendance. There was pressure on the President to have a chief guest who was distinguished to impress the members attending.

My son Brynnen was studying at St Thomas's Hospital in London to become a doctor. He became friendly with a fellow student, Miss Nicola Griffiths, whose father was Deputy Chairman and Managing Director of Sainsbury's at the time. However, in his early career Mr Griffiths had had links with the mining industry. When he graduated from the university, during the Second World War, he was given a choice: he could either join the Secret Service or he could become a Bevin Boy. Due to his background, he chose to be a Bevin Boy in the Staffordshire coalfield. Many months before the date of the annual dinner, I asked Bryn to approach Mr Griffiths to see if he would be the main speaker at the event. As a sort of bribe, I personalised the request with the offer that he and his wife could stay as our guests at Vissitt Manor and I would take them on a visit to the Selby Coalfield. He accepted the request and locked it in his diary. Then began a niggling worry for me: my son had a reputation for playing the field with his girlfriends. If his association with Nicola Griffiths ceased, I would probably lose my chief guest at the dinner! Fortunately, that didn't happen, although later he did change his girlfriend to a nursing sister at St Thomas's, who later became his wife.

Roy Griffiths and his wife were charming people, and a delight to host at Vissitt Manor. He was an enthusiast for his supermarket and, in talking to us, explained the intensive efforts being made to introduce computer control systems into the stores. Those systems became the backbone of the checkouts and the information for stock control and reordering.

The visit to the mine had to be changed from Wistow, due to a dispute with the workforce, and we went to Riccall mine. We descended through the shaft in the sinking hoppitt into the pit bottom. Our wives were with us and there were elaborate arrangements of ladders to get them into the hoppitt and out of it. Roy was interested in the mine as it showed the massive excavation and support girders for the pit bottom and the way they keyed into the shaft walls. The contractor's shaft sinking workers were amazed to see two ladies as visitors; that was a first for them and they were surprisingly keen and gentlemanly to help them around the pit bottom and then back into the hoppitt. Later we lunched in Wistow House and met other Selby Coalfield staff.

That evening at the dinner, in Sheffield, there was some mumbling among the membership.

'Why has he invited a grocer as his chief guest?' 'What does he know about mining?'

When Roy Griffiths started speaking he related his experience in the Staffordshire pits, and it soon became clear that he knew quite a lot about coal mining. The theme of his speech was to focus on the common problem shared by coal mines and supermarkets — they both had logistical challenges moving the products to the point of use underground and point-of-sale in the stores. He was a good speaker and the members were very impressed.

Much later, after he had left Sainsbury's, Roy Griffiths was invited by Mrs Thatcher to do an investigation into the Health Service. His initial impression was of major shock, in that nobody seemed to know what any individual procedure, or treatment, or prescription cost. He certainly knew what everything cost in his stores and what the margins of profitability were in all their operations.

TREVOR MASSEY

## Changes at Vissitt Manor

In 1983 Fred Hopkinson who worked the farm which surrounded Vissitt Manor decided to retire and put up the farm for sale by auction. It was bought by two adjacent farmers and the land divided by them. The field next to the Manor house, which had been used for grazing shire horses, was changed to arable crops. The septic tank for the Manor House was exposed within the arable field. That farmer had no use for the stables and we were able to buy them from him, along with a strip of land around our eastern boundary. We adapted the L shaped stables into a rectangular format so that we had a very large building for stocking equipment and produce. In the new land our nephew Robin installed a new septic tank and connected the out-flow to the existing drainage system.

The new land was filled with trees to form, over a few years, an effective wind break from the prevailing westerly winds. There were no crops grown on the new land but it had a vital role in the garden operations which will be described later.

Two other structures were installed around this time. A conservatory was built onto the house adjacent to the back door, which did reduce the wind chill into the kitchen side of the house. A 20 feet diameter Soladorm greenhouse was built in a central area of the garden which captured maximum sunlight. This became highly productive and its use will also be described later.

Clear plans had now been made on the layout of the gardens. The vegetable areas would be in four feet wide blocks separated by paved paths. There would be no edges to any grass areas so that the grass cutting machines could run on the paving stones to cut the edges of the grass. In the front garden of the house paving bricks were cemented in place to achieve this.

In the greenhouse loose bricks were laid to create a two-foot wide bed around the outer edge of the building. Two D shaped beds were laid out either side of a central path through the greenhouse. All these bricks were dismantled in November and stacked up where they would get purged and cleaned by frost and rain. All the soil was emptied out and used for fertilising the potato areas for the next year. The bricks were rebuilt and the whole greenhouse filled with fresh compost at the beginning of March.

## The first coal face at Wistow

The first strip was taken off Wistow A1's face on 27 June 1983, and over the next four weeks performance grew. The coal was loaded out and dispatched to the local power stations. It has an average ash content of around 12% which was marginally too low for the Central Electricity Generating Board; but suggested that the concept of producing a saleable product without any treatment other than blending might be feasible.

However, on Saturday 23 July, water from the Permian strata broke through at A1's face and a very severe 'weight' affected the middle third of the coalface and damaged many of the powered support legs. Fortunately, the water flowed off the coal face and out along the face roadways. A new roadway had been developed, several hundred metres long, to the dip side of the Wistow main roadways and it was decided to forfeit that roadway temporarily, with its equipment, to act as a sump until pumping systems were installed in the pit bottom with additional ranges up the shaft. The water rate was measured at 120 to 150 litres per second. Wonders were achieved by the engineers to build tanks in the pit bottom, install large capacity pumps and pipe ranges in the shafts. When the new system was commissioned the water was pumped into a convenient dyke which ran alongside the Wistow site.

Work was done to repair the supports on the coal face and operations restarted. With commitment from the men, it was agreed to keep the face moving on a continuous basis over seven days to avoid the possibility of shear breaks developing during the weekend stands. Production restarted on 1 September. This pattern continued, confidence increased, and on week ending 29 October, the best output figure of 27,633 tonnes was obtained from A1's face. The onset of the miners' overtime ban introduced intermittent working, and further serious weights were experienced, causing damage to the chocks, an increase in water flow and further recovery on the coal face was necessary.

It was a similar story with A2's face. There were very high performances but, despite changes to the design from A1's face, intermittent weights and inflows of water affected operations.

In view of the experience on A1's and A2's it was decided to design a new mining system (see below).

The early coal faces demonstrated some special characteristics. The faces were virtually free from methane gas, which is unusual for the Barnsley seam. The strata above the seam were very strong and this must have contributed to the roof breaks reaching up to the Permian strata, 80 metres above. There was obviously bridging of the limestone and bunter sandstone, as there was little subsidence on the surface. The deeper coal faces at Wistow didn't have the weight and water problems of A1 and A2 faces.

## Selby during the miners' strike

The onset of the full miners' strike presented a challenge to running the Selby Coalfield. There were numerous contracts throughout the project that were operated by non-NUM workers. The aim was to keep this work progressing while accepting that Selby management, like the rest of the industry, would have to keep the pits safe without the help of any NUM men. Two police forces had responsibility for different parts of the coalfield: West Yorkshire police and North Yorkshire police. There were NUM pickets active from the start of the strike. I rang both police forces requesting their presence to ensure the pickets were not preventing non-NUM workers getting to their work. I got a very lukewarm response, as the police clearly had no intention of getting involved in the strike at that stage.

During week two of the strike I had to attend a meeting at Hobart House in London. Unusually, I went by car. Before the meeting, I spoke to one of my colleagues at Headquarters and explained that we, the management in the coalfields, did not have clear instructions from the Board as to how we were to operate during the strike. His response surprised me. He said that the Board were not clear at that stage what the government position was on the strike.

After my meeting, when I went down into the basement to my car to travel back to Yorkshire, my driver told me: 'Mr McGregor has been to see Mrs Thatcher this morning'. Drivers and senior PAs are always the source of much strategic information!

The die had been cast by the government. When I got back home, the police were ringing me up to see what I needed from them. The police chose Gascoigne Wood to hold their resources for controlling the pickets around the Selby Coalfield. On several occasions I drove up to the Gascoigne site behind the mounted police team. I had no idea of the size of police horses until I looked up at them from the car.

Plans were proposed in some parts of the country to get NUM men to return to work with special transport arrangements, as the strike was failing. At Selby, we got involved in this initiative near the end, but it was on a small scale. There was one occasion when I attended a lunchtime meeting with local authority officers at the Monk Fryston hotel. I was told that I had just missed a secret meeting between Arthur Scargill and Mr McGregor. My meeting made some progress, but there was no indication that the secret meeting had brought about any positive results.

The 1984/1985 miners' strike gave a slot of opportunity to allow the design and manufacture of the coal face equipment to operate short, 45 metres long, single-entry faces at Wistow. This system was approved by the Mines Inspectorate, who were involved with the design. The concept required the application of a special ventilation system which was automatically monitored and allowed the short face to move back quickly along its single roadway. The single-entry short coal faces were applied after the end of the miners' strike. They were operated by a small team of three men and one official, who evolved a sawtooth system doing two passes for a full cycle. This meant that the shearer did not do any shuffling at either end of the face, it just cut back to the other end. There was therefore a continual flow of coal from the face, which gave around 20,000 tons per week. The maximum weekly tonnage from a single-entry face was almost

27,000 tons, during a week in June 1987. It placed high pressure on the mine to develop the short faces quickly and to move the equipment efficiently from one coal face to the next. It was clearly an interim arrangement until standard retreat coal faces could be developed further to the east of the Wistow reserves. Those faces were brought into production after I had left the coalfield and they had dry conditions.

## World Mining Congress, India, November 1984

Two weeks before the 1984 World Mining Congress, Mrs Indira Gandhi was murdered while walking in her garden. She was shot by two of her guards, Satwart Singh and Beant Singh. This was a major international incident and there was initially the possibility that the Mining Congress would be cancelled, however, the Indian Government quickly issued confirmation that it would take place. With the miners' strike affecting staff throughout the mining industry, the NCB revised its list of the people who would attend. Mike Eaton was to present a paper about the Selby Coalfield Project, but he had been appointed to London to carry out a role as Coal Board spokesman during the strike. I was to stand in for him at the World Congress and there were other senior staff who were replaced. The detailed arrangements for flights and accommodation were coordinated by the Head of Method Study for the Coal Board, who had experience of working in India.

It was a remarkable few days for the delegates. The Congress was very well organised by the Indian team, for both the technical presentations and the social events. But there was also time for us to see the way of life in Delhi. We were staying in the Taj Mahal hotel, on one of the higher floors. We looked down on the fog of polluted air being breathed by the local population. We saw history being made when we witnessed the many thousands of people who had attended the presentation by Rajiv Gandhi, as they were returning from the open-air event. At one social event, the whole grounds of a large hotel had been decorated and there were elephants and other artists performing for our entertainment. We also saw the gap between the rich and the poor in India: some two-foot diameter concrete tubes laid outside the hotel for a major drainage project, were being used as sleeping quarters by the very poor; yet within the hotel we witnessed one or two weddings where the bride and guests were weighed down with gold ornaments.

We took a bus tour to Phatipur Sikri and the Taj Mahal. At that time, the local feeling against the Sikh community for the assassination of Mrs Gandhi was clearly on display. Our bus driver was a Sikh. At one point there appeared to be a hold-up in the traffic flow. He immediately left the road and speeded past the stoppage. Phatipur Sikri is a remarkable construction of buildings which are left unoccupied, as the planners had failed to identify a source of water for the community. The Taj Mahal is a building in a class of its own, in world terms, which changes in appearance as the different light sources affect it. One or two hard-baked technical representatives on our trip were emotionally moved by what we saw. That was an indication of something really being world-shattering!

Giving the paper about the Selby Coalfield was not an unusual role for me and I have no recollections of any slip-ups with the visual aids, or being presented with difficult questions.

One memory of that visit to India which persists, was a dinner given on the final evening by one of the mining companies. It was well organised, in a private room, but the menu included chicken which, on inspection, seemed distinctly undercooked. Messages were signalled by some of our wives to be wary. Those who didn't eat the chicken were unaffected, but the rest of the company was subject to a severe dose of Delhi-belly which persisted on the plane journey home. On the flight home, Jack Wood announced that he had decided to change his name; in future he would be called Ken Wood, the human liquidiser!

### An interim evaluation

Would the Selby Coalfield achieve its targets? This was a question raised by many interested parties within British Coal, and by others in the coal industry. In one of our reports to the Coal Board we said that it would indeed achieve its objectives, and in fact we considered the coalfield potential was higher than the predicted 10 million tons per year. When all the mines were in full production we suggested it would be possible to exceed 11 million tons annual output. In 1993-1994, that potential became a reality when the output from Selby exceeded 12 million tons.

However, there were opportunities to simplify the layout. During the negotiations for planning consent, the position of the Whitemoor mine site was moved so that it was nearer to Riccall. This posed the question of whether the coalfield could have operated with four satellite mine sites instead of five? Before I left the coalfield, plans were made to connect Whitemoor to Riccall and to connect North Selby to Stillingfleet. These link-ups would allow the length of the Spine Roads to be reduced to 12,600 metres from the original 15,000 metres.

After I left, as a component of the privatised UK Coal, the success of the coal face design was demonstrated when 240,000 tons was produced in one week from one coal face at Wistow — a remarkable achievement.

One concept that was not achieved, though, was to provide a coal product for the market without any coal preparation facilities. The early planners for the coalfield had postulated that it might be necessary to import dirt to achieve the desired ash content for the power stations. The early coal from Wistow gave a saleable product slightly cleaner than required by the power stations. However, where the Barnsley seam at the other mines was thinner than at Wistow, the dirt produced in driving the main roadways increased the dirt proportion in the run of mine and the saleable product needed some treatment. Within the covered stockyard, a barrel washer was installed to scalp off sufficient dirt to give a product with an ash content suitable for the power stations. Planning permission was given to form the extracted dirt into a tip on a site adjacent to Gascoigne Wood. Over the years this dirt formed a heap that appeared as an interesting feature in the flat countryside of the Vale of York.

In total, 121 million tons was produced during the life of the Coalfield and there were no significant problems of subsidence or unacceptable environmental issues affecting the Vale of York.

So why was the coalfield closed? In 1998, the selling price of the Selby coal was reduced by 20%, a challenge that few industries could face, even with the best technology. The coal industry was contracting at that time and the requirement was for less mines, but the ones which survived had to be highly productive. Selby was hampered by its sheer size — it couldn't be slimmed down. It was not viable at 5 million tons per year and the coalfield was losing £30m per year at that level of output. Selby was a double-decker bus of a mine, but it was useless as a mini-bus or a taxi. There were ample coal reserves available in the coalfield, but an application to the local authority to grant planning permission to work the coal east of the river Derwent was turned down. It was not challenged or pursued by UK Coal, who had probably concluded that any additional reserves could not be worked profitably. So, the coalfield was closed in the financial year 2003-2004.

### Two 'After Eight Mints' from Wistow

There was one team of men recruited for Wistow who had previously worked for a Mining Contracting Company doing work for the Coal Board. They were an excellent team who achieved good performances driving a trunk roadway at Wistow. I met them on several occasions on my visits underground. Their objective was to try to get on an enhanced rate of pay, like the one they had had with the contracting company. We had some lively discussions on the subject, but we couldn't agree to their requests. Their charge man knew this would be the outcome, but one of his team was a persistent arguer whenever we met.

Then one day, when I visited them, the arguer remained silent and ignored me. I asked the charge man if he was ill, but I was told he had abandoned the battle as he never won anything from me.

On my next visit to the team, I was accompanying a VIP visitor to the mine, along with his wife. My wife Enid was also in the party, as support. The two ladies caused the usual impact on an underground visit, and the workmen were suitably friendly and refrained from using expletives in their language. The arguer among the workmen suddenly realised which lady was my wife. He immediately found his voice.

'Come over here, Luv, and let me tell you what he is really like!'

Enid has never told me what he said to her!

When the Wistow manpower was nearing its full complement, an issue arose affecting the pithead baths. The manager decided to use the downcast shaft to bring the men to the surface at the end of their shifts. The large cage could bring the total shift manpower on a single wind of 170 men. His reasoning was that this removed the normal problems at most mines of pushing and shoving to try to get on the first draw to the surface. However, it conflicted with the planned flow through the pithead baths which was based on the men riding up the upcast shaft which would take several winds. The architect approached us with a plan to take up additional land and extend the bathhouse building and increase the number of showers in the baths — quite a job and quite a cost. As discipline in the pit bottom at shift ends was a problem at many pits, I supported the manager's decision.

I decided to visit the baths and see for myself the size of the problem. I found the baths attendant and took him around with me to get his opinions. The showers were different to normal designs: some were single shower units, but the majority were communal blocks with four shower heads along a length of the shower area. The baths attendant said he thought there was a simple solution to the problem.

'If I make all the shower heads in the communal units into double shower heads, there should be plenty of room for the lads to have a good wash. Our lads are not shy, they'll not grumble if they're a bit crowded, so long as they have their own shower head.'

So much for the expert solving a problem — the man on the job often knows the right answer!

### Senior Management moves post-strike

Within British Coal, as it was renamed, there were numerous senior management changes after the strike. At Board level, John Northard was appointed as Production Director and Ken Moses as Technical Director. They had both been Area Directors before the strike, and Ken Moses had been noted for his very positive role in organising the workmen to return to work in the North Derbyshire Area.

Their different roles were clearly defined. John Northard, through the Area Directors, covered the production of the coal; Ken Moses had the role of managing all the technical services in the industry, including mining research, approving capital expenditure schemes, overseeing the Purchasing and Stores Department, and the Management of the Central Workshops. Ken also oversaw Compower, a separate computer business that covered all the Board operations and carried out contracts with some private companies.

The two appointments were appropriate for the people appointed. John Northard was 'a detail man', who revelled in getting to the bottom of the mining problems in the collieries. Ken Moses was a strategic thinker who was always seeking change within the industry to match the changing circumstances the business faced. He was keen to influence his staff to think strategically 'outside the box'.

# Chapter 17

*A move to Headquarters*
*1985-1991*

One day, Ken Moses approached me at an Institution of Mining Engineers meeting and suggested I should move to Headquarters, as Head of Mining, to work in his Technical Department. I would be stationed at Bretby, near Burton-on-Trent, which was the centre for mining research. I told him to speak to Enid who was also at the meeting! She was supportive of the proposal, but made it clear that we would not move from Vissitt Manor in Yorkshire.

The initial idea was that I would get a flat near Bretby, to live in during the week, and return to Yorkshire at the weekend. I did a few visits to Bretby before my official start date, and on one of them I had a different driver, who did the 70 miles between Bretby and Vissitt Manor in under 70 minutes. This raised the option of travelling each day. It proved to be a wise decision because on most weeks I had commitments away from Bretby, at meetings in London, in the coalfields, in Europe and occasionally overseas. I got a reliable driver, who lived in Castleford, and we left my home at 6.30am each morning to be in my office at Bretby before 8.00am, and we usually arrived back at Vissitt Manor in the evening around 6.30pm. It was a long day, but I used the travelling time in the car as thinking time. Near the end of my time at Bretby, we got a car phone, and this helped as I could leave the office before 5.00pm and take phone calls as we travelled. In later years, with the increased density of traffic on the motorways, such regular home-to-office journeys would not be a practical proposition.

### Technical Department organisation

The three key senior staff working for Ken Moses were myself, as Head of Mining, Colin Williams, as Head of Supply and Contracts Department, and Martin Shelton, as Head of Compower. It became normal, as relationships developed, for the four of us to meet and discuss the strategic issues facing the Corporation. There were times when Ken would introduce major conceptual proposals that needed a thorough review and modification. The three of us developed an arrangement whereby we didn't criticise his initial ideas, but we would agree to apply ourselves to review how the proposals could be applied. Ken was prepared to accept modifications to his basic concepts, if the arguments were well presented, and this was a characteristic of great value to his key staff.

It became clear that Ken's relationship with John Northard was not easy, as their styles of management were very different. John Longdon was John Northard's second-in-command and I had a good relationship with him, so there were numerous issues that John Longdon and myself solved without them going up to the Moses/Northard level.

### First impressions

The development of the Bretby headquarters had been pioneered by Peter Tregellis, fully supported by the Coal Board. He believed that the industry needed a centre of excellence developing new technology and techniques for improving the efficiency of the industry. He ensured facilities were available at Bretby to educate the various levels of management in the industry and to explain the opportunities offered by the new technology. The Bretby site was in many ways like an industrial university. Edwina Curry, the local MP, boasted that she had more PhDs working in her constituency than most other constituencies in the UK.

It is difficult to describe the diversity of the staff and their different roles. Some were pure researchers working on specific research projects; others were expert at lecturing and selling the results of research and development projects to the operational management in the field; there was a group doing 'blue sky' research in evaluating new technical ideas and R&D from around the world, and considering what could be applied in British Coal.

The site had a suite of lecture rooms which could host meetings for major lectures and conferences, along with smaller rooms for break-out meetings. There were excellent catering arrangements for daytime and evening functions and there was a large demonstration hall with an overhead crane that could be used for the build-up of new coal face equipment and other new mining systems.

There was another nearby site, at Swadlincote, that was used for some practical research in simulated coal face conditions. It was also a large enough site to allow demonstrations of new coal face machines and powered supports in operation, working a synthetic coal seam that was constructed for a range of seam sections.

Bretby was placed centrally in the UK to allow access from all the coalfields in the country.

The organisation chart for Bretby (on the next page) was devised along with Ken Moses in my early days at Headquarters. It is complete except for one additional member of staff who was drafted in from another department — Deborah White effectively covered administration and Colin Ambler concentrated on staff issues. Deborah was a fiery character, who feared no one, and could be relied upon to focus everybody's efforts to achieve the objectives of the department.

**C T Massey**
**Head of Mining**

### ENVIRONMENT AND PLANNING
Head of Environment and Planning
D J Stephenson

**MINE ACTIVITY FUNCTIONS**
- Manager, Surface Environment (Production)
  E J Allett
- Manager, Mine Environment
  W Highton
- Manager, Safety and Legislation
  M Widdas
- Manager, Project Appraisal
  D Slatcher
- Head of Business Planning
  K Mordue
- Chief Method Study Engineer
  I J Watson

**PROFESSIONAL SERVICES**
- Chief Surveyor and Minerals Manager
  A H Sturgess
- Chief Geologist
  M J Allen

### ENGINEERING
Head of Engineering
W J Alexander

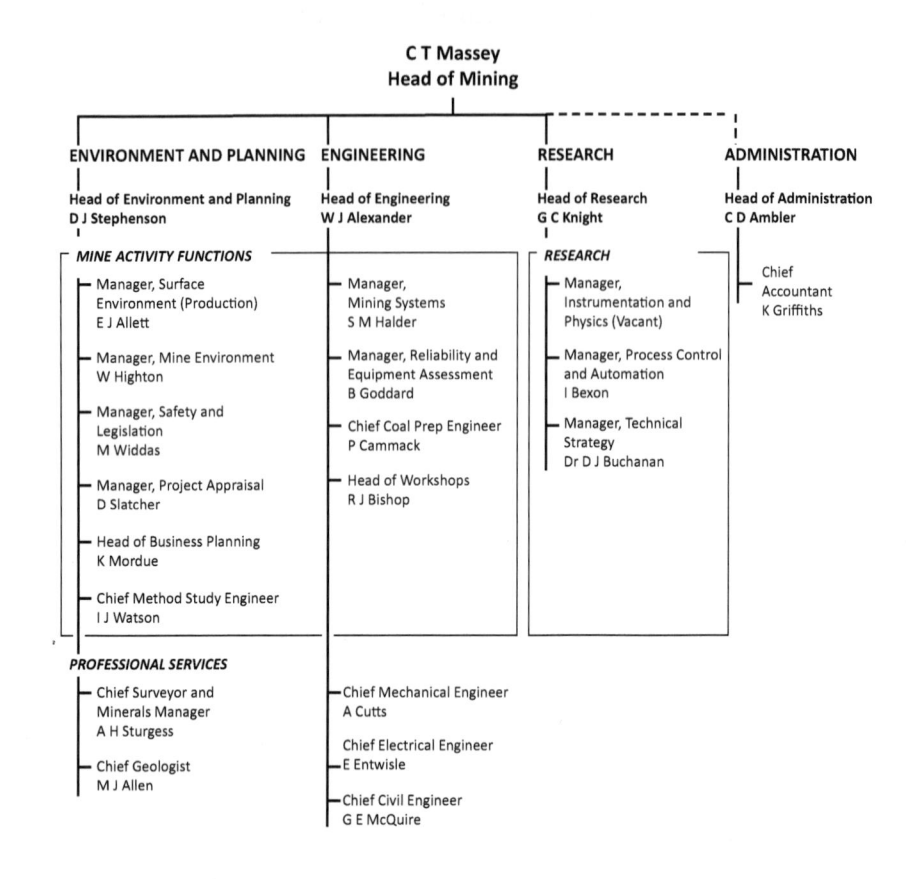

- Manager, Mining Systems
  S M Halder
- Manager, Reliability and Equipment Assessment
  B Goddard
- Chief Coal Prep Engineer
  P Cammack
- Head of Workshops
  R J Bishop

- Chief Mechanical Engineer
  A Cutts
- Chief Electrical Engineer
  E Entwisle
- Chief Civil Engineer
  G E McQuire

### RESEARCH
Head of Research
G C Knight

**RESEARCH**
- Manager, Instrumentation and Physics (Vacant)
- Manager, Process Control and Automation
  I Bexon
- Manager, Technical Strategy
  Dr D J Buchanan

### ADMINISTRATION
Head of Administration
C D Ambler

- Chief Accountant
  K Griffiths

*Organisation plan for Bretby*

## Queen's Award for Industry

Very early in my time at Bretby, we received notification that we had won an industry award for developments of control systems for auxiliary ventilation fans. Shortly afterwards came the invitations to a reception at Buckingham Palace for the people involved in the research work. The invitation was to the head of the organisation, a senior research person who had worked on the project, and a craft employee who had worked on the equipment. As the work for this award had all been done before my time at Bretby, I referred it to Ken Moses for advice. His reaction was that I should lead the team at the Palace and if I didn't go he would go.

It was a chance in a lifetime for the research staff member, but particularly so for the craft operator. I wanted it to be memorable for all of us. We stayed overnight in a hotel and I arranged dinner for us after the Palace reception. The views in Buckingham Palace were truly amazing and the event was organised with military precision. Each group was presented to the Queen, who was supported by the Duke of Edinburgh, and then there were other senior political and business dignitaries circulating in an open forum.

The occasion was made special for us by an intervention by the Duke of Edinburgh. I introduced my colleagues to the Queen and the Duke after explaining where we came from in the Coal Board. I then assumed that we should move on and let the next group introduce themselves. But the Duke would not have this, 'Go on, tell us how it all works.' The research expert readily gave an answer to the Duke's questions. On subsequent meetings within the Royal Academy of Engineering I later saw further examples of the Duke's inquisitive approach to engineering subjects.

It was a successful diversion for me and an example of the excellent work done at Bretby. It was a significant and memorable event for my colleagues.

## The Bretby senior management team

The team under me was constituted of very experienced people within the industry, which we assembled at Bretby soon after I took up my appointment.

| | |
|---|---|
| Head of Research | Geoff Knight |
| Head of Engineering | Bill Alexander |
| Major Projects | Derek Stephenson |
| Staff Manager | Colin Ambler |
| Finance | Keith Griffiths |
| Head of Administration | Deborah White |

It was this team, which worked with me during my time as Head of Mining, and later as Head of Technical department, to run the organisation to increase its efficiently. We then had to adapt the organisation to be appropriate to the changed needs of the Coal Board.

Initially, I had the PA of my predecessor, but she opted for promotion to another post at Bretby. I then got a new PA, Jenny Smith, who was a truly amazing person, with skill and expertise respected by everyone in the Technical Department. When the organisation was fully established, I was asked by Ken Moses who I wanted to nominate to stand in for me when I was away on holiday. I had given this some thought, as there was a number of the senior staff working for me who were all of equal status. If I nominated one, there would be quite a few people who would be disappointed. After much thought I favoured an alternative solution, which I decided to put to Ken Moses. I had some uncertainty about his reaction. Jenny Smith was so good at her job and she understood so well how the whole organisation worked that, in my absence, I felt that she would know the best person in the organisation to deal with any issue that arose. On this basis I didn't intend to nominate any of the Managers to stand in for me — Jenny would be in charge. Ken Moses had such a great respect for Jenny that he agreed to my proposal.

PAs are a key resource in any large organisation. As a pro-active PA, the way they plan the diary can achieve 50% additional output from their boss compared with a re-active PA. Jenny was one of the very best and she had a steely forcefulness which she could apply, if necessary, in dealing with staff on behalf of her boss. There was one technical advantage in the organisation of the secretarial system at Bretby: the secretaries of the senior staff had their computers linked on a closed circuit that no one else could access.

On one occasion, I made a proposal that didn't suit Jenny. When we were preparing complex papers, which had to go the Board of British Coal, on occasions Geoff Knight would offer to make a first draft to speed up the preparation of the papers. He was a good drafter, so I usually accepted his offer. It dawned on me that Geoff must have a terminal on the secure link of the PAs' computer system. If Geoff had a link, then why couldn't I have one? When I suggested this to Jenny her face immediately fell, and I knew I was treading on holy ground. It was clear that she thought I was sure to muck up the system. Her steely look was underscored as she quietly insisted that she was quite capable of providing me with any papers I needed. On reflection, I thought her worries might be well-founded. Jenny had recourse to a supporter. I had a visit from Deborah White, the Head of Administration, with an official reprimand. She said it would be quite inappropriate for me to have a terminal linked into the PAs' computer system. I had to wait until I retired to dabble my feet into the mystery of using a computer for word processing.

The annual budget for the department was over £7million pounds, which included the spending on R&D projects, some of which were financed by EU grants.

### Concentrating the staff at Bretby

The first priority was to bring together all the staff of the different disciplines onto the Bretby site. There was a team in London who had worked for the Director General of Mining, whose role had disappeared. They had been there for a good number of years and were reluctant to leave London. However, discussions took place with them and transfers were arranged; they were able to settle in the country areas around Bretby. Once they were established with their families, they revelled in the locality and the country facilities, and they confessed that their initial worries about the relocation were unfounded.

There was an office in Doncaster which housed the engineering disciplines and the specialists for coal face operations, safety and environmental engineers. We were not very successful in persuading some of the engineering staff to relocate to Bretby. For a month or two only one mechanical engineer relocated to Bretby. He had to be assisted in his work by the Chief Mechanical Engineer and the Head of Engineering until new appointments were made.

The Bretby environment was a challenge for a lot of staff who we brought into the department from the operations in the coalfields. The intensity of day to day work in the mines, with different geological, operational, and human relations issues continually interrupting the planned continuity of the operations, seemed far removed from the semi-academic life in the research fields at Bretby.

Many of the research staff regarded me as a production man and they were wary that I might change the way of life they had experienced over the last few years at Bretby. I knew from my early experience in looking and listening and questioning, that it was inevitable that there would have to be changes throughout Bretby. It was also becoming clear that the Coal Board would face significant changes as a result of the miners' strike.

## Early plans for change

There were new techniques available to increase efficiency in the mines. New, heavy duty, machines were being introduced for cutting the coal on the coal faces. These were backed up by heavy duty panzer conveyors on the coal face. Conveying the coal from the coal face to the pit bottom needed ever-bigger conveyors and there was the possibility of automating them. That technology had been designed and was available from Bretby. Similarly, there were more heavy-duty machines available for driving the mine roadways. The application of these machines would allow more coal to be produced by retreat coal faces.

The other challenge was to monitor the environment underground in the mines so that the risk of explosions or fires was eliminated. The technology of remotely monitoring the environment was also available at Bretby. So, an initial challenge was to improve the application of successful research projects into the mines. But were all the programmes of research realistic, and could they satisfy the current needs of the industry?

Geoff Knight had one short session each week with one of the different research teams to try to address this question. The team leader, with a colleague, would make a brief presentation on the work they were doing and then answer questions from Geoff. I went to some of these sessions, when I could, and joined in the discussion. In many cases the ideas and the research being done was to be admired, but its relevance, and timescale for release into the industry, was not clearly defined.

However, one of the legitimate frustrations for research staff was that the application of successful developments of systems and equipment which had undergone successful trials in the mines, was not being exploited fast enough. Was there any way to speed up the applications?

After discussions, we decided to create Activity Teams for different systems. The teams would be made up of research specialists with members of our different technical staff. They would 'sell' the concepts to the mines and assist to make sure that more applications were installed in the mines and they stayed to ensure they worked successfully. This gave the research teams the satisfaction of seeing their work applied in the field. However in some cases, after a number of successful applications, the novelty of seeing their work applied in the field lost its lustre and they wanted to get back to their R&D role.

## Christmas traditions at Bretby

One of the traditions at Bretby was to hold Christmas lunches for everyone on our books. This had to be spread over two days, when nearly 200 people were crammed into an extended dining room and served four courses of the Christmas menu. It was a major achievement in cooking and logistics for our chief cook, Doreen, and her staff. The occasions were held in the festive spirit to celebrate the closing year, anticipate the Christmas holiday break and ponder the challenges of the coming year.

My role was to attend both lunches and give a 'state of the union' speech about the coal industry and our work in Technical Department. This was a major challenge. In 1987, bearing in mind the effects of the year-long miners' strike and the political repercussions and publicity surrounding the coal industry, the audience were keen to hear what I had to say. In the early years the aim was to try to give some reassurance that the industry was settling down and had a future. In my later years there were some results of output and productivity which were very positive. There were also continuing successes, and more Queen's Awards to celebrate for our research work.

How were these better results achieved? The National Board had introduced an incentive scheme for staff, as a complement to the incentive scheme for coal face operations at the mines. For senior staff to achieve a bonus of 20% of salary the objectives were set at meteoric levels of productivity increases, and they were initially discounted as completely unachievable. Over time, with the wider application of heavy duty coal faces and the reduction of manpower through the pit closure programme, the meteoric levels of productivity were achieved.

However, the fundamental prospects for British Coal were not good, as the price of international coal on the world markets was below the production costs of the Coal Board. The result was that the international port of Immingham, which was developed to handle coal for export, had its role reversed, and coal from South America and the USA arrived there and was dispatched by train to the large power stations in Yorkshire and Nottingham.

I had help from my senior staff in preparing these Christmas speeches and I usually had the odd humorous item, which, including good wishes for the season, aimed to send the audience away with a smile on their faces.

## *Examples of R&D challenges*

For generations, one of the challenges in the mining industry had been the impact of dust produced in the mining operations and its effect on the workmen. Pneumoconiosis had been identified as an industrial disease which affected all miners. A more serious version was silicosis, affecting men working in mining operations where they worked in places with rocks with a high concentration of silica in them.

The Institute of Occupational Medicine in Edinburgh had been involved in the coal mines for years, measuring the dust levels in different parts of the mine. Their conclusion was that pneumoconiosis was directly related to the level of dust in the air streams of the workings. Systematic samples were taken of the dust levels throughout the mines. Also, miners had a chest x-ray every five years, and their lungs were evaluated for any change over that period. If they had deteriorated over the five years they were given a progressive index; if the progressive index was not increasing it showed that the dust suppression in their working conditions was succeeding.

I served on a committee of the Institute of Occupational Medicine reviewing the experiences in the coal industry, and I attended quarterly meetings in Edinburgh. At that time a member of the committee was Sir Richard Doll, who had proved the link between smoking and lung cancer. He was a charming man, and his experience was helpful in reviewing the results in the coal industry.

It was clear that the new heavy-duty shearers on the coal face would be cutting much more coal in any shift than before, and would be producing larger quantities of the fine dust which did the damage to the miners' lungs. Dust suppression on the shearers had been through high pressure water sprays directed onto the cutter picks on the shearer discs. Each coal face was systematically sampled to measure the dust produced during a shift in the return airway of the coal face. There was some indication in these samples that the new higher horsepower machines were producing more dust than was acceptable to give conditions such that the progression indexes shown by the x-rays did not increase. Was there any possibility of devising a new system of dust suppression that would solve the problem?

One hazard that had arisen with the shearers operating where there was a sandstone roof, was for the cutter picks to strike the sandstone roof and produce sparks. Very occasionally a spark would ignite any methane that was collecting in the roof near the cutting horizon. A solution to that hazard was to produce a jet of air, along with the water, onto the cutter picks that diluted any methane around the picks. If it was possible to blow air onto the cutter picks, would it not be possible to suck air and dust from the cutter picks? This was the concept of the dust extraction shearer drums. The central tube which had been used to feed air to the picks was used as the route for air and dust to be extracted from the cutter picks. The air and dust were deposited onto the panzer conveyor with the coal. The technology of fitting the suction unit and the piping into the discs was complex, but it worked and the dust concentrations on the coal faces were significantly reduced.

The ongoing objective was to design that technology into the smaller diameter shearer discs for the thinner coal seam sections. As the trials were conducted with dust extraction drums designed for thinner seam sections, the technology proved to be successful. I was always briefed on these successes and the team involved were thrilled to be extending a solution to the health hazard of dust in the production faces of the mines.

Mining is a undoubtedly a dangerous operation. It has a very long history of major tragedies killing many men. The danger of explosions due to methane being released from the shale beds adjacent to the coal measures, which were disturbed and fractured when the coal was mined, was a continual challenge to mining engineers. The development of equipment which was intrinsically safe, and would not give a spark or flame to ignite any accumulation of methane, was one major element in reducing the risk of explosions. Also, techniques of methane drainage were applied to drill the strata and capture the methane into pipes and transfer it to the surface to keep it from the mine ventilation system. It was often used for heating on the colliery surfaces, as a cost saving exercise. However, the methane was not produced as a constant stream, and its access into the mine ventilation was variable. It was a silent, invisible, odourless gas that could appear into the mine workings without any of the workmen or officials being aware of its presence at dangerous levels. Was it possible to have an automatic monitoring system that would indicate abnormal circumstances?

The 'blue sky' research team at Bretby suggested that pellistors should be investigated for application in the coal mines. A pellistor is a solid-state device that will measure the presence of a combustible gas in an air stream. The ones used in the instruments applied in the coal industry indicated very small changes in the methane content in the air at the place where they were deployed. The conversion of the readings into a digital format allowed the information to be transferred around the mine and then to a surface control room.

The technology was developed at Bretby, and software was evolved, so that a pit could have an environmental package for the whole mine. Every coal face would be monitored, and other key places in the mine, where the air streams were combined, such that an overall environmental picture of the mine was visible. The software also allowed settings to be included so that warnings were given if levels were exceeded in any location. The package was released to the industry under the name MINOS (Mine Operating System). Staff from Bretby were deployed to the mines to assist in the installations and to brief colliery staff about the technology. From time to time the software for the MINOS system would be updated and tried at one mine. If it proved successful, the software could be applied at all the installations in British Coal.

Investments to introduce these packages into the long-life mines gave an immediate safety payback. As they were installed they secured the end of any explosions in British Coal mines — from that time forward to the end of deep mines in the industry!

## *Legislation*

The coal industry had been dominated by the various Coal Mines Acts and Regulations for 150 years. These were admired throughout the world and had been used by other industries to develop their own legislation packages. All mining engineers, other engineers and specialist staff were schooled in the detail of this legislation, which defined how the mines had to be operated and how they had to be managed. There was scope to operate with an exemption to some parts of the legislation if agreed with the Mines Inspectorate. However, when the industry was undergoing major changes in the equipment and the methods of working over a short timescale, it was not feasible to wait for new legislation to be covered in the Acts or Regulations. An alternative was required.

The Coal Board had evolved a system of Technical Department Instructions (TDIs). These were developed by appropriate experts from the Technical Department consulting with senior people in the industry, with the Mines Inspectorate, and in some cases with companies manufacturing equipment for the industry. When the documents were finalised, they would be approved by the Technical Director and issued throughout British Coal and to other relevant bodies. They were mandatory and had to be applied throughout the industry in the same way as the Coal Mines Acts and Regulations.

It was an efficient way of ensuring that the technical standards in the industry were maintained and enhanced. It was not easy to get agreement with all the interested parties on any particular issue and, at times, there had to be a bit of arm twisting to make progress. But TDIs allowed the industry to move forward applying new technology.

## *Other people's problems*

We were delighted in the 1980s in British Coal not to have any major incidents which made the headlines with multiple deaths. We systematically studied other industry's major incidents. Our safety team specialists were required to crawl through the facts of any major incident and assess their relevance to the coal industry. One such incident was the King's Cross station fire on the wooden elevator. A fire underground is a major disaster in a coal mine and has resulted in some mines having to be sealed off and the pit closed.

There were features of the King's Cross fire that stimulated questions and research to try to explain how the fire suddenly surged up the elevator. This was explained after tests at the Safety in Mines Research Establishment in Derbyshire, where it was concluded that the flame was searching for oxygen when it swept forward along the upper section of the elevator. However, it was agreed that the King's Cross fire was started by a cigarette falling onto a mixture of grease and dust under the elevator. That conclusion was quickly reported by our safety team. It could not have happened in one of our mines for two reasons: firstly, everyone was searched for cigarettes and contraband before they went underground (it was automatic dismissal for any person caught); secondly, only non-flammable grease was used on all our underground equipment. Such information underlines the pay back for having good practice and insisting on the right specification of products.

## An evaluation by the Boss

The organisation was finalised, and the staff were in place. How was it working? Ken Moses decided to do his own review. He discussed it with me, and his plan was to have a couple of sessions with team leaders of all the different technical groups. His aim was to get answers to the following questions: 'What is your team doing; what is the priority for the different objectives of your work; what effect will the results of your work have on costs for British Coal.'

We discussed this within the senior staff team at Bretby. There were legitimate concerns on two counts: firstly, evaluating the financial pay back for some of our teams was difficult to assess as their work was to supply a technical service and expertise to the mines. The success of their work was often measured in problems that had been avoided, which is difficult to evaluate in financial terms. In other cases, the application of modified machines, which had enhanced reliability, showed an immense pay back that could be measured financially.

The second problem was that some of our staff were unfamiliar with the role of making presentations and selecting the key elements of their work that would highlight the best impacts on the Coal Board costs.

We decided to tackle the issues by having a rehearsal with each of the presenters. We knew that Ken Moses would be forthright with his questions and would destroy some of the team leaders unless we gave them guidance of what to say and how to say it. With the critical support of my senior staff, we worked through the presentations. Some of them required major editing and textual revisions; others needed much more rehearsal by the presenter so that they spoke the words with confidence; yet others needed improved visual aids to improve the clarity of the information. There were quite a few re-runs and I was hopeful that the final presentations would give Ken Moses a view of our work that he could accept and support. I knew that he wanted a positive story to take to the British Coal Board meetings.

However we made one serious mistake. One of our staff, who was responsible for machinery developments, had qualifications and wide experience in public speaking. We knew he would make an impressive presentation, so we virtually skipped his rehearsal. On the day of the presentations, I had to leave before the end for some meeting elsewhere. As I departed, I was reasonably confident that the Bretby staff were putting over a good picture of our work in the Technical Department and there had been no major issues raised by Ken Moses. The next morning, I learned that there had been a major outburst by Ken. While his presentation was impressive, our man on machine developments, had used a slide which showed the distribution of his section's work. It clearly illustrated that the major effort was on new machines and only a smaller percentage on designing improvements to the existing machines and equipment. Ken spotted this and ridiculed the priority of the section's work. The opportunity to affect the Board's results required an upgrade of our existing fleet, as they would be in a majority for a long time. As a result of the critical questions and the force of Ken Moses' attack, our man was shattered and destroyed. He resigned from his post shortly afterwards. It was a tragedy and it was the fault of all our senior team as we had not been diligent to review every presentation in sufficient detail to spot an inappropriate figure on one slide.

## Representation on external committees

It was to be expected that the Coal Board would be represented on numerous Technical Committees pertinent to the coal industry in the UK, in Europe and internationally. The attendance at these committees fell mainly on Ken Moses and myself, but due to his commitments as Technical Director of the Board, I often had to appear on his behalf. The following is a list of the major committees and what their role was.

### ACCORD Committee

This was a major UK government committee established to monitor all the nationalised industries and evaluate their research programmes. Its membership included senior civil servants, academics and representatives of coal, electric, gas, nuclear, rail and the post office. The format of the meetings was often a visit to a research establishment with presentations and demonstrations of the work they were doing. Each year the committee reviewed the research programmes of the different industries. They were very critical if the financial figures showed a reduction in spending for any member industry. This was the case for the Coal Board as we trimmed our research expenditure as the industry was declining significantly in size. One year a first draft of our programme, tabled by the committee, was rejected by Ken Moses. He insisted on the Chairman of the committee, with the Secretary, staying on and working with him to redraft the conclusions of the paper.

Ken was a member of this committee, but I took his place on several occasions.

I recall two interesting visits that I made in his place. At the Post Office research centre, in a presentation, we were shown how they were evaluating the possibility of introducing postcodes to accurately define locations. Postcodes were applied in the UK and have now become standard technology used by many organisations in carrying out their business.

I was present at the ACCORD meeting and demonstration at the Torness nuclear site in Scotland. I was surprised to find that they had planning permission to build more nuclear generators on the site if they wanted. It was also surprising to see that they had built a control room for training purposes, at a cost of many millions of pounds, even though there was one available in England. The cost of the one nuclear plant at Torness was around £1.6 billion — the same as the total cost of the development of the Selby Coalfield.

At that time, the government was promoting nuclear plants as providers of the cheapest electricity into the National Grid. When I wrote my report to Ken Moses I quoted figures given to us at Torness and concluded that there was no way that nuclear generation could be producing the cheapest electricity. It was only when electricity generation was being evaluated for privatisation that the market quickly made its mind up that nuclear was not a viable risk for investment. The true cost of nuclear generation was redefined and increased, still without any costs for dismantling the nuclear stations at the end of their life. As a result, the nuclear stations were not included in the privatisation process.

## EU Committee on Coal Research

Membership of this committee became my responsibility, but Ken Moses wanted a report on all our meetings.

From January 1986, Italy and Spain joined the European States, giving a total of 12 members. The committee met each quarter to review the coal industries in the various member states and the progress with their research programmes. Once each year there was a meeting to agree the annual allocation of funds and approve which research applications should be supported for each country. Most countries applied for more funds than were available and there was hard bargaining for the final allocations.

Collaboration between the largest coal mining countries — Germany, France, UK and Netherlands — took place to agree priorities for the individual submissions. However, the EU permanent officials were keen that the money should be shared out on a broad basis. As a result, support was given to projects from the likes of Portugal, to do research work that had been done years previously in the UK and Germany — not a very efficient way of spending investment funds. In general, in the UK, we got support for our priority projects, but we didn't get back the amount of cash the UK had contributed to the EU coal research funds.

Coal mines were being closed in the European coal industries in the late 1980s, and there was interest in comparing the progress being made in the different countries. I attended one meeting and reported that in the UK we had reduced the manpower in British Coal by nearly 20% in the last six months. There were gasps from the German delegates at that statement (I liked to try to make the Germans smile in our discussions. On this issue I said I might be able to arrange a transfer of Mr McGregor, our Chairman, to them, if it would be helpful. They were too shocked to smile.)

The French policy on mine closures was to delay closure until new jobs were provided in that region for all the men at the mine. With such a policy, in many cases, the progress with restructuring was delayed by many months. There was also no policy to adapt the mine sites for other use, as applied in the UK. Many mine sites in regions of the UK were worked for opencast output and the pit heaps profiled so that all signs of a coal mine were removed.

For one meeting, the agenda was very challenging in the time available. The German Chairman requested that the meeting took place in English, which would avoid the use of interpreters. A French member, who was so adept at English that he frequently corrected the interpreters, said he could not accept the suggestion as he could not understand my Yorkshire accent!

These meetings usually culminated in me hiding in a corner of Brussels airport using my dictaphone to record the outcome of the meeting for Jenny to type up and circulate to Ken Moses and colleagues in Technical Department.

## International Committee on Coal Research

This committee met annually in the different countries. Every third year there was a conference organised by the committee, which attracted delegates from around the world. The key member countries were USA, Canada, Australia, New Zealand, Japan, UK, France and Germany. I was a member of the committee and succeeded John Mills, a Board Member of British Coal. My first attendance was to a meeting at St Louis, USA, where I gave a paper outlining our policy of 'activity teams' to stimulate the application of successful R&D projects into the mines.

Subsequent meetings around the world included an underground visit to a mine in the Hunter Valley in Australia. It was unusual to be given a white boiler suit, worn over our night clothes, as our mining outfit; it was equally surprising to find that the first miner I spoke to came from Grimethorpe, within two miles of our home at Vissitt Manor! Another visit to an opencast mine clearly showed the low-cost production available in parts of Australia. There was only around 4 metres of shale cover above the seam, the seam was 6 metres thick and it was worked by a machine operated by one man which cut the coal and loaded it directly onto an extendable conveyor at a rate of up to 600 tons per hour. Nothing was operating on the site, though, as it was subject to a wage dispute!

In Canada, we were flown up to the Athabasca tar sands centred on Fort McMurray in Alberta. The logistics of moving the large volumes of the tar sands to the treatment plant was impressive. More exceptional was the briefing on the operating conditions in the winter, when the sub-zero temperatures required that the dump trucks had to be kept running all the time, because if they stopped it wasn't possible to re-start them. The massive treatment plant handling all the sands resulted in a flow of oil into a pipe which went underground to the south. The massive reserves are viable when the price of oil is in its higher ranges.

One year the annual meeting was in Japan. There we were introduced to a completely different way of life. The ladies were escorted around Tokyo and into the home of one of the committee members. The role and responsibilities for the wife contrasted markedly with those of wives living in Western democracies, who were the majority of the ladies attending. There was a final dinner for all the delegates and their wives. It was a sumptuous occasion in the Japanese style. We sat on the floor and were fed numerous courses by beautiful and charming young ladies. It was difficult to identify what the food was, or whether it was raw or cooked. The same applied to the different drinks. What is memorable, is the finesse with which the young ladies presented everything to each person as though they, and the food, or drinks, were very special. Unfortunately, the Chairman had requested that, after the dinner, several items not concluded in the meeting should be discussed. It went on for quite a while and the beautiful young ladies were retained in a role of waiting on the guests. They had never witnessed before such a situation where they were present to serve but what they would do was not defined.

It was the responsibility of the UK to host the annual meeting in 1992, my last year with British Coal. I decided that we would hold the meetings in York, rather than London. We arranged a visit to the Selby Coalfield, which was of interest to the mining engineers on the committee. There was a ladies group, as usual, at the meeting, and I arranged for a friend of mine, who lived near York, to lead the ladies in a walk around York looking at the historical buildings. He knew of gems around the town which were rarely seen by visitors, who usually just flocked to the Minster. He was also a brilliant communicator, who could fascinate children, as well as adults, on a range of subjects. He used his charm on the ladies, particularly the American wives, to give them a unique historical experience that they would never forget.

We were also able to hold the final dinner at an exceptional hotel on the outskirts of York racecourse. The ambience of the private dining room and the food was greatly appreciated by the Chairman and the committee members along with their wives.

It had not been an easy gathering to organise, as the guests arrived from all parts of the world by air and rail, spread out over a significant timescale. We had my secretary, with an assistant, and two cars and drivers to get the members to the hotel for the start of the event. At the end of the three days, on the morning after the final dinner, there was a similar logistical challenge getting everyone to the railway station or airports to continue their travel plans. Some guests were taking the opportunity to enjoy a few days in the UK before returning to their home country. I received a letter of sincere appreciation from the American Chairman of the committee for the way the whole event had been organised by our team. I showed this letter to my staff, and duly praised them for the way they had worked together to make a success of this special international event.

*Safety in Mines Research & Testing Site, Buxton, Derbyshire.*

I also served on the Advisory Board for this site. The site had done excellent work in testing explosives and other systems and equipment for the mining industry. Open days on the site were used to demonstrate good practice to workmen and officials in the coal industry. It had an area on its large site to demonstrate how a methane explosion could be propelled into a much bigger coal dust explosion if the coal dust was not suppressed by being mixed with stone dust. These demonstrations left a lasting image with all observers of what can go wrong in coal mines if coal dust is not suppressed.

During my time on the Advisory Board, the laboratories and staff were taking on work covering other industries besides coal, for the Health and Safety Executive. In time, the site was renamed the Health and Safety Laboratory Buxton. It was another example of technical expertise developed in the coal industry being used to establish safe working practices in other industries.

*Nuclear Waste*

There had been investigations over the years to try to find an underground site in the UK where nuclear waste could be stored safely for the long term. The geology had to be exactly appropriate and there had to be no risk of geological activity affecting the underground conditions. With the closure of mines taking place it crossed a few minds that there might be an appropriate mine closing that could give underground access to the right geology.

I was approached by two organisations and asked: 'Are there any closed or closing mines which are suitable to exploit for nuclear waste?' They both explained that they would be able to employ the redundant coal miners to do all the mining work, creating the chambers to hold the waste containers. I regarded this last proposal with some reservations.

The examinations by our chief geologist, surprisingly, gave positive answers. There was one colliery in the Midlands coalfields, which was scheduled to close in the near future, where the mine shafts were just over a mile from a major granite intrusion. In Cheshire there was a closed mine, where the shafts had not yet been capped, that could provide access to the salt measures. So, there were two sites, with geological rocks compatible with nuclear waste storage, that could be considered.

I met with Ken Moses and briefed him on the information, as it would require a Coal Board decision if we were to get involved. Not only was the possible employment of redundant miners a sensitive industrial relations issue, but there was a possibility of serious community objections to having a nuclear waste project in their locality. British Coal would likely be severely criticised for making the sites available by any community affected by a nuclear waste site. Ken was clear that it was a hostage to fortune for the Coal Board to get involved in any proposals relative to nuclear waste. Whether he discussed it at Board level, I am not sure, but I reported to the organisations involved that we did not, at that time, have any mines giving access to suitable strata of the required geological reliability for underground nuclear waste storage. That was thirty years ago (1987) and the search still goes on for a suitable site.

### *The right organisation for the coal industry*

The coal industry was drastically smaller than it had been on nationalisation, and the organisation had been simplified. The Divisions had disappeared, and the number of Areas was reduced. Could there be opportunities for further changes? Ken Moses was considering this issue, and was seeking experience from elsewhere, even internationally, for guidance. He decided to investigate how a major coal company in the USA was structured. He chose Peabody, which was one of the biggest coal companies in the USA. He sent Colin Williams and myself to visit Peabody's new headquarters in Henderson, Kentucky. We travelled to the locality and stayed overnight.

We were at the headquarters first thing in the morning and had meetings with the senior company staff. It quickly became clear that the company operated with a small management team at the mines, but that all the various technical expertise and staff were held centrally and deployed to the mines as and when required. We were given details of how this worked, and the numbers of people involved. Peabody staff, who knew the full extent of the company organisation, with management numbers and disciplines, were made available to us. They willingly gave us detailed statistical information about the company. It was a long day, but when we got back to the hotel for the night we had a record of the complete management numbers and the different technical disciplines involved, and we also knew where they were all located throughout the whole of the Peabody Company.

The outcome of this fact-finding mission was that in British Coal, mines were formed into Groups with a tight management structure, and specialist engineers were held at headquarters. Some Group Directors clung on to the experts they had known and worked with in the Areas, but there was a simplification of the organisation. This placed a clear, enhanced role for the Technical Department at Bretby. Most of the staff in Technical Department accepted that they had responsibilities to support the Group operations on any technical problems and good relationships were established.

I recall two examples of this in action. In the safety team we had one or two experts at dealing with spontaneous combustion incidents underground. There was a difficult incident of heating at a mine in

Scotland, which was active over the New Year holiday. In the end they found the source of the heating and put it out. The key expert came back to Bretby and reported that solving the heating was not the main problem — it was finding a handful of men prepared to go down the mine on New Year's Eve!

The same men were also experts at investigating household explosions. Several times each year there were gas explosions which destroyed property and caused injuries. Where did the gas come from? We immediately investigated these incidents as the Coal Board could be involved. There have been cases where a mine has been closed and methane has accumulated in the old workings and risen towards the surface. The methane can then sometimes travel through certain porous rocks to collect in the foundations of nearby properties. More often, though, the gas has been generated in waste disposal dumps which have not been drained of the methane they produce. In these cases, the gas migrates into houses near the dumps. Our team, with old mine plans and information about local waste disposal dumps, could usually identify very quickly where the methane had originated. It then allowed remedial actions to be undertaken to safeguard other properties in that locality.

## *An additional role*

In the original formation of the National Coal Board, there had been a Board Member for Science. His operational responsibilities covered the laboratories which were in place around the country. These laboratories gave an analytical service to evaluate all the statutory safety samples taken throughout the mines. The samples included methane ventilation samples, dust samples and water samples. Sometimes there were one-off environmental samples taken to try to solve technical problems. Also, there were samples taken of all the different coal qualities sold to customers. These results were used to achieve the selling price of the product. Again, detailed samples of the saleable coal would be taken when a new coal seam was brought into production; similarly, a new or modified process in a coal preparation plant would require detailed sampling to ensure the saleable coal was to the required quality.

Some laboratories had been closed and their work load concentrated at other laboratories. However, a review of the forward requirements indicated that all the workload could be handled by one central site. As we had a suitable building that could be converted, Bretby was chosen to be the site. It took nearly a year to effect the change and was the responsibility of one of the staff who covered the maintenance of the Bretby buildings. He did a brilliant job, as the technical and safety requirements of a laboratory are very challenging. He kept me informed of the progress with the building conversion and then the installation of the intricate sampling equipment. I visited the site with him a couple of times to see the progress, but it was only when I was leaving Bretby that I noticed on his lapel that he wore a small clip showing that he was a member of the Methodist Church; I should have got to know him better and earlier.

When the laboratory was commissioned my title was changed from Head of Mining to Head of Technical Department.

### Slimming down the staff

There had been some progress with staff saving when mining department staff was concentrated at Bretby. However, with the closure of numerous mines it was clear to me that we had to review the staffing levels throughout the Technical Department. I was aware that we had some very able, and highly qualified staff; I also knew that British Rail, Rolls Royce and other major companies around Derby, were recruiting research staff. There were also the first rumours spreading of a possible Toyota development at Burniston, which was quite near Bretby. Might it be harder for the Technical Department staff to get suitable employment in two years' time? I thought that was likely to be the case, so it was right to devise a plan.

Ken Moses made a helpful move when he organised a dinner, followed by talks, with the union for management staff — British Association of

Colliery Management (BACM). The dinner was held in a private room in a hotel in Nottingham. It was not a confrontational meeting; the senior officials of BACM knew the position that the Board was in, and that there had to be some staff reductions. Ken gave them assurances that all staff would be treated fairly and fully briefed on what was taking place. Colin Ambler was there from Bretby and it would be his job to manage the process. There were no discussions of likely numbers involved, or how the selection would be organised. We left the meeting satisfied that we were all working towards the same objective. On our way home, our drivers told us that there had been a serious plane crash in Scotland. That evening, 21 December 1988, Pan Am flight 103 was blown out of the skies with a bomb and 270 people, some from the community of Lockerbie, lost their lives. It made our discussions pale into insignificance compared to the impact on the people affected directly and indirectly by the Lockerbie disaster.

The staff at Bretby were expecting some changes, but they were unlikely to anticipate the staff reductions I thought were necessary. The New Year of 1989 was the time for action. We needed an organisation with a staffing level that could be in place for some time into the future. It was not sensible to go through all the stress of a review which only made limited reductions, and then need to repeat the exercise in twelve months' time. How was the plan to be made and how could it be kept secret until it was approved by the Coal Board?

There was one advantage in the organisation of the secretarial system at Bretby. I was confident that the secretaries of the senior staff were all dedicated and responsible and they would ensure that the records of our deliberations on staffing levels remained confidential. We needed to agree the complement of each section of our organisation. Our aim was to agree the numbers, not the names — that would come later. There were some sections that we thought had a lower rating in priority in the revised circumstances of British Coal. So, the levels of staff reductions varied around the Department. We also were aware that some of our staff, for a variety of reasons, would welcome the opportunity to take redundancy.

It was clear that the atmosphere in the Technical Department was rife with rumours of what was to happen. I was approached by one senior manager suggesting that we should not proceed with our reorganisation as it was upsetting the staff. In my view there was nothing to be gained by any delay. A paper was prepared explaining the plan, and identifying the staffing implications, and it was passed on to Ken Moses. It represented a reduction of staff by over 20%. He accepted the plan and considered it might encourage other departments in the Corporation Headquarters to consider slimming down their staffing levels as well.

With approval from the Board, we moved to the next stage. Each manager was interviewed by one of our senior staff and told what his complement was to be. It would be the Manager's responsibility to decide which staff he wanted to retain. He was given only two days to make his mind up and he was instructed, in the strongest possible terms, not to speak to anyone about the figures. We then constructed a list of all the people on the retained list and this became the Monday list, as that would be the day of the week that it would be released. All the other names would be on the Thursday list, as that would be the day when the persons to be declared redundant would be finalised.

The names list was issued on the Monday. Everyone on the Monday list had the security of a job. But they were told that they had the option to be made redundant. To opt for redundancy, they would have to make their mind up in 48 hours, so that they could be replaced by someone on the Thursday list.

The mechanism of the Monday and Thursday lists might be criticised for being tight on time, but I wanted to get the staff settled down as soon as possible so that we could get them all focussed on the work we had to do in the Department. The system worked, and there were some staff on the Monday list who wanted redundancy, and there were some on the Thursday list, who were glad to remain working.

The redundancy terms on offer from British Coal were good, and Colin Ambler, Head of Staff in Technical Department, tried to arrange to handle those leaving to their convenience. There was one case that I handled directly. British Coal had a 'Blue Book' defining expenses that could be claimed by anyone who had to travel around the country to do their job. One or two of our staff knew the details of the expenses schedule in the finest detail, and had exploited it over the years. It could involve staying overnight instead of making a late journey home. When I saw the expenses claim of one member of staff, I instructed that he had to spend his remaining weeks on our books in an office in Doncaster and not travel anywhere. Someone else would cover any work scheduled for him and we would show a significant cost saving with not having to pay his expenses.

It is a lonely life being head of a Department undergoing a major slimming process. There are no set ways to carry out the exercise; it has to be approached step by step and the moves forward have to be agreed with the senior team. But the head of the organisation has to take the lead and hold his nerve, despite the fears and rumours circulating around the establishment. There is no way the man in charge can pass out to his staff the responsibility for making decisions about the livelihoods and careers of the staff, without his full involvement.

## An approach from a senior member of staff

Bill Alexander, the Head of Engineering, ask to see me on a confidential matter. He was a tough character, who pushed all his engineering staff to give of their very best and his aim was to lift the engineering standards in the Coal industry to the highest possible level. He was doing a good job and I respected him as a foremost member of the senior management team at Bretby. His request was a surprise: he had been offered a senior appointment with Thames Water. He wished to take the offer, but he could not afford to accept it unless he received favourable severance terms from British Coal, because of the very high housing costs in London. Would I approach Ken Moses on his behalf, to see if he could be released? I agreed to discuss it with Ken Moses when I had given it some thought. I didn't want to lose Bill, but I also didn't want to stand in the way of promotion for any of my staff. My deliberations required me to think who could replace Bill. The possible replacement I had in mind gave some hope for Bill getting what he wanted.

I met Ken Moses and explained Bill's request. Ken's immediate reaction was 'no way', but I went on to explain that, as far as I was concerned, the critical decision was who would take Bill's place as Head of Engineering. Having reviewed who was available elsewhere in British Coal, the one who I thought I could work with was Tony Caunt, who had worked with Ken Moses in North Derbyshire Area. On the basis of my proposal, Ken Moses was supportive of letting Bill go.

It was not the end of Bill Alexander as far as I was concerned. We were to meet from time to time in the future at Engineering Institution meetings, and subsequently at Royal Academy of Engineering events. Bill joined Thames Water at the right time. A major project was in operation to drive a ring main around London to improve the efficiency of water distribution for the capital. The project was behind schedule and had technical problems when he arrived. Thames Water senior staff had a 'hands-off' approach to major projects and contracts. Their view was the company had bid for the job, so it was their responsibility to sort out the problems. If the job was delayed, it would cost the contractor financial penalties. Bill took with him the Coal Board philosophy that if there were problems with any contract it was the responsibility of everyone involved to sort out the problems to get the job done. Bill's 'hands-on' approach led to a modification of the tunnelling equipment and the way the job was organised. The result was that the ring main zoomed around London and was finished early and under the specified cost. Bill's involvement with the problems of the ring main and the solutions was obvious within Thames Water. He was rightly proud — his qualities had been displayed. He moved into senior management roles and was eventually appointed Chief Executive of the company. In that position he was faced with the full exposure of the media on operational and environmental issues. He handled those elements of his new post in the same direct way he tackled engineering challenges. Bill Alexander was elected as a Fellow of the Royal Academy of Engineering, and also served on their Council while he was at Thames Water.

Through Ken Moses' initiative I was elected to be a Fellow of the Royal

Academy of Engineering in 1990, and became involved in efforts to inspire young people to choose an engineering career. The Royal Academy ran a 'Headstart Programme' through a number of university engineering departments. This was a one-week intensive experience at a university, undertaking lectures and practical laboratory work, to get the flavour of an engineering degree course. It was taken at the end of the first year in the sixth form, before the students had selected their university courses. Some university staff involved in operating the Headstart courses regarded this group of highly motivated students as a gift presented to them to recruit for their own engineering courses. After I retired, I became a member of a small team of Fellows who visited the various courses to evaluate their impact and effectiveness. We were highly critical of the universities who sought to recruit the Headstart students to their own degree courses as the aim was to inspire students to choose an engineering career. The best options for university courses varied depending on which engineering subject was chosen. We also pressed for a part of each course to include a visit off-site, to an engineering firm, to see engineering in action. I recall one such visit on a Birmingham university Headstart course, to the Cadbury factory to witness the manufacture and packaging of their famous chocolates. There was some interest in the chemistry of making the basic chocolate, but the engineering involved in the wrapping and packaging of the boxes of chocolates was very impressive!

It was very satisfying to have time with these young people, and reassuring to know that they could become leading engineers of the future in the UK.

## A board meeting at Bretby

Since its formation on nationalisation, the Coal Board had met for all its meetings at Hobart House in London. Robert Haslem, its Chairman, decided that it should have some of its meetings out in the coalfields. Whether it was Ken Moses who offered, or someone else who suggested us as a suitable victim for being the starter, I was informed that Bretby was to be the host of the first Board meeting outside London. This was to be a challenge to everyone at Bretby and it was that challenge that was emphasised as the information was passed out to all our staff.

We had a good story to tell, and we had the facilities on the site to display it. We set up the demonstration hall with a range of equipment and systems that the Board members could visit. The selection of the key staff who would explain what was on display, and answer any questions, was handled by Jeff Knight, Head of Research. He had some enthusiastic staff who could sell their products. One such person was Wendy English, who covered the MINOS Systems for environmental monitoring. She had a display which could simulate the monitoring system in action and show how the software operated to ensure that the mine environment stayed safe.

There was another display which showed a shearer which was equipped with a monitoring and control system which lifted and lowered the cutting disc so that it stayed in the coal seam. This technology reduced the amount of dirt produced from some manually operated shearers.

There was a display which showed how a dust extraction drum worked and reduced the dust produced by the new heavy-duty machines.

The aim was to fill the Demonstration Hall with ideas and systems which demonstrated the range of our work in all aspects of mining coal. Deliberately there was too much for the Board members to be able to see everything, but we were confident that they would be surprised when they saw some things that they had no idea that we were involved in.

The schedule was that the Board members would arrive in the locality to stay in a hotel, for those not travelling from home the following day. There was an evening meal at a hotel, which we hosted, and I had the job of welcoming them and explaining the arrangements for the following day.

It was a logistics nightmare, with Board members arriving from distant parts of the country and from overseas. Organising the transport rested with Deborah White, who had a team of drivers and cars at her disposal. I recall on the following morning, when we had got everyone onto the Bretby site and into the room set up for the Board meeting, she suddenly realised that the team of drivers had had no time for breakfast. She got on the phone to the manageress of the dining rooms, who was a prima donna. In her most strident tones, Deborah tackled the problem. 'Doreen, I am sending a dozen drivers across to you now who have had no breakfast. I want them feeding with no questions asked.'

## A new mine challenge

Throughout my time in technical department I tried to fit in visits to mines in different parts of the country. There was a new mine, Ashfordby, being developed to work quite a large coalfield in the Vale of Belvoir. I was intrigued to see this development, to compare the technical approach to that of the Selby Coalfield. It was a single mine development, aimed at two million tons per year. The shafts were complete when I visited the mine, and roadways were being driven out from the pit bottom. I was rather shocked to see what had been decided about the roadway support system. To compensate for the anticipated difficult strata conditions, a decision had been made to drive circular roadways. These would be supported with segments which locked together to give a complete shield, similar to a support system behind the tunnel boring machine at Selby. However, they were driving the roadways with a standard road header and there was no mechanised system to erect the supports. The rate of progress was thus very poor. Also, much to my dismay, one shaft had been equipped with special cages to handle the segments for the circular roadways. They had thus lost having that shaft, with a very large cage and counterweight, which could handle complete machines and heavy built up pieces of equipment into the mine. It was always disappointing to see that successful systems in one part of British Coal had not been applied elsewhere. The cage and counterweight had proved a very efficient system at Selby.

By the time of my visit I think that there were doubts being expressed locally that the wide application of circular roadways in Ashfordby was going to be successful. An alternative technology was required. There was, fortunately, a new technology for roadway supports already being considered in British Coal.

## *Roof bolting*

The standard supports for roadways in British Coal mines had been passive supports in wood, or later with steel arches. The strength of these support systems, even with heavy duty steel arches, had sometimes proved inadequate for the strata stresses which arose when the coal was extracted.

An alternative support system, used particularly in American mines, was applied where the method of extracting the coal was by 'bord and pillar' mining. The support system was to apply roof bolts for the roadways used by their continuous miners. Essentially, bolts were drilled up into the roof, often through a strap, and tensioned to lock the strata beds together. The pillars initially left provided support for the mining system. However, the Americans were adept in their lay-outs at robbing a proportion of these pillars. They did it by using their continuous miners to cut into the pillars without setting any supports as they retreated from any block of coal.

Bearing in mind that there were no provisions in any parts of the Coal Mines Act or Regulations to allow the exclusive use of roof bolts in any mining system, the opportunities for any roof bolting in British Coal mines was not straightforward. Roof bolts had been used in circumstances where passive supports had failed, and repair work was necessary. So, some mines and miners had experience of how roof bolts could lock strata together. Was there any hope that roof bolts could be used in British Coal as primary supports? This was a challenge for the mining engineers in the Technical Department. There was much discussion internally, and references to overseas experience were considered.

A mining engineer returned from a visit to coal mines in Australia and reported that he had met a young man there who was a leading expert on roof bolting techniques. His name was Winton Gale. I spoke to Ken Moses about this and proposed that we contacted Winton Gale and give him a contract to come to the UK and visit some of our mines and give us his assessment of the possibility of applying roof bolts in our mining conditions. He agreed to come!

Winton was a young man, with a very quiet manner, who had acquired extensive experience in Australian mines with variable conditions. We arranged for him to visit various mines and meet some of our senior staff. I was keen to see how his quiet, confident style would be received in the coalfields. Obviously Ashfordby and Selby were included in his initial tour. Albert Tuke, the Area Director for the North Yorkshire Area and Selby, was a very strong personality, who had been pressing for new technology to boost the performance at Selby.

When Winton had completed his tour around some UK mines, we had a debriefing meeting with him. He said that he could design roof bolting systems that would be able to offer safe support for a range of the mining operations he had seen. He explained that he had designed monitoring which could be applied to the roof bolts to measure the stress on the bolts as operations progressed. It was really a traffic light system which gave a green light when there was no significant stress on the bolts, but it changed to an orange light when the stress was building up. When the stress on the bolts was so high that the bolts could fail, it showed as a red light. This monitoring system meant that the bolting pattern in any application could be assessed on scientific measurement, not on chance. The initial feeling in British Coal was that failure would be defined by a roof fall. Winton Gale's proposals were much more positive as far as the Mines Inspectorate and the trade unions were concerned.

Winton Gale returned to Australia for a short spell to have an input into his business there. We set up with him another, more extensive, contract for work in the UK. He would design roof bolting support systems in a range of sites. In each case he would work with colliery staff alongside Technical Department staff to ensure that the bolts were drilled and installed correctly. The sites would be monitored to ensure they gave the specified support to the strata. It was clear that there was a significant learning curve to spreading this technology around British Coal. Wherever Winton Gale worked in the industry, he had impressed management and men by his quiet confidence and expertise in the introduction of roof bolting schemes.

We decided that we must expose this Australian widely in the industry. To do this we set up a one-day conference at Bretby, where Winton would have a free hand to demonstrate the extent of his experience on roof bolting in a range of different mining conditions. (He had already done some work in other countries, as well as Australia and the UK.)

We invited the designated engineers on roof bolting from every Group of British Coal. Also, we invited representatives of the NUM and NACODS trade unions from every Group in the country. I rang the Deputy Chief Inspector of mines and invited him to the meeting. He said he would come but he wanted all the Mines Inspectors who had responsibility for any roof bolting proposals anywhere in the country to also be there.

Wow! What an audience. I had everyone that mattered on this new technology — the workmen, the trade unions, the Inspectorate, the local experts and the Technical Department experts. What I'd hoped would come out of this day was an agreement to press forward with more applications. But we also needed an agreement that all the schemes would be governed by a new Technical Department Instruction prescribing the detailed standards which would apply to every installation. These objectives rested on the shoulders of a quiet Australian who had to sell his vision and expertise to a very diverse audience. When I introduced Winton, I wondered if it was a bridge too far to achieve so much from one meeting. However, Winton Gale did not let me down; he was brilliant, and everyone was convinced that we could proceed with assured, safe systems. Sometimes, having all the eggs in one basket may be a risky gamble but, if successful, it can be a major step forward and in this case it allowed the exploitation of new technology much faster.

Work started straightaway to prepare a TDI to cover roof bolting. From that time the new technology began to be applied more widely in British Coal, with more applications of roof bolting being adopted.

I went back to Ashfordby to see the roof bolting on roadways there. There was a large increase in the rate of progress in opening out the new roadways, and a growing confidence that the roof bolting was giving stabilised roof conditions.

## The Toyota car plant at Burniston

The Toyota development became a reality and a few of our staff obtained posts on the project. After some time, we got feelers about a possible use of some of our Bretby facilities by Toyota. This was a puzzle, and it took a while before the request became explicit. What they wanted was the use of our large Demonstration Hall. It was a strange request, as we were still using it from time to time. It became clear that they saw the Hall, with its overhead crane, and access for motor vehicles, as an excellent stocking site. They wanted to use it to hold strip steel if 'just in time' failed! I'd expected that Japanese management would have had more confidence in the provisioning of its components to manage its operations, without such a significant insurance policy!

Nothing came of the request, as there were other options on the horizon. It was not the end of my contacts with the Toyota plant, however. A few years later, after I'd retired, I joined a party from the Royal Academy of Engineering on a visit to the site. It was a memorable experience.

We were briefed by the UK manager of Toyota. He explained that they were starting to export the Toyota Avensis model from Burniston to Japan. To satisfy that market, they had initiated upgrades on the car to satisfy the requirements of the Japanese male drivers. Japanese men often lived in small sized houses, where there was no designated space for the man of the house, so the car was effectively his private space, and he wanted it to be to the highest possible specification. Some men even had a separate pair of shoes just for driving the car.

When we toured the production line we saw that no man had any buttons or zips evident on his clothes that might mark the bodywork. The standard of cleanliness throughout the whole factory was higher than would apply in any hospital ward or operating theatre. Each car was built to satisfy a specific order, so they were all in different colours. The body work was covered in cloths so that no dirt could possibly get to any parts of the new car. At one point, the doors left the car and went on a separate production line to be built up; they reappeared at the right time to be fitted to the car.

The tools and implements used by the men on the line were impressive. There were no signs of panic or pressure on the line and I got the impression that quite a few of the young men had been trained in the coal industry to get qualifications to enhance their engineering background.

The factory could not in any way be compared with a UK engineering works of the 1960s. It was truly magnificent.

### *Capital Investment Committee*

An important responsibility for Ken Moses as Technical Director was approving all capital projects in British Coal, working through the Capital Investment Committee. Ken was a thorough chairman at these meetings and aimed to gain a realistic evaluation of each project by questioning in detail the official presenting the project and then eliciting contributions by the experts in attendance. He held a monthly meeting to manage this and it was regarded as an important meeting in the Board's business. Papers were submitted for each project and these were assessed by the different departments — finance, production, technical department, supply and contracts department and marketing department. It was a large committee and I was the member for Technical Department, along with Colin Williams for Supply and Contracts Department.

There was a lot of pressure from the Areas, and subsequently Groups, to get their projects through. In Technical Department we had a lot of work to do as we had to assess projects over our full range of specialisms — engineering aspects, safety and health in the mines, environment impact, geological risks, subsidence potential and impact on local safety and health in the community.

A major requirement in the submissions was an assessment of what the project would deliver as an improvement in the financial results in the Area or Group. Some of the claims were not realistic. Our aim in Technical Department was to try to resolve issues on the facts provided, the assumptions made, and results projected, with the people submitting the project, rather than have them exposed in the meeting. It was not always possible to do this, though, and there were occasions when projects failed to get approval and there was bad blood between those submitting the project and headquarters staff assessing it.

On one occasion, a project in Wales was put forward with claims that it would give a very significant improvement in the Group's current performance. The figures used for the current performance had only been achieved over a few weeks just before the submission. I pointed this out in the discussion and, after the meeting, I was attacked as a 'shit' by the Group Director for questioning the reliability of his figures, even though the project got approval.

The Capital Committee also noted the completion reports submitted for all the capital schemes agreed by the Board. Accountability for the success or failure of these schemes was covered by the accountability meetings held with the different Areas or Groups or Departments within the Corporation. Like any large organisation, British Coal suffered with movements of staff such that new faces were often around to answer for the success or failure of schemes which had been managed over several years.

## *Another idea to exploit Bretby's facilities*

British Coal had a large site at Chalfont St Giles which operated as a Staff College. Most management staff had attended various courses there, and it was used for seminars for senior Board staff at critical times in the industry. It had also introduced a specialist, six-weeks long, Senior Staff Management Course for potential high flyers in British Coal. It used international lecturers and tutors to achieve a top quality international rating. It replaced the former practice of sending senior staff away on international management courses. I had been on the second of these senior staff courses at Chalfont, which ran through June 1973, to 2 August 1973. I was then a production manager in South Yorkshire Area. Ken Moses had done a spell as a senior tutor at Chalfont during his career and, after he had moved on, he was a frequent guest speaker there in their programmes.

One day, I suggested to Ken that at Bretby we had every facility that a staff college needed, except overnight accommodation. That would be easily provided if we constructed accommodation units in the upper level of a part of the Demonstration Hall which connected directly to the lecture theatres and facilities for conferences. This would allow the sale of the Chalfont site, which must have a significant sale price in that part of the country. Ken said he didn't feel that it would be possible to get the Board to agree to make the move at that time. So, the idea was not dead, but it was kept confidential to me and Ken Moses.

A few months later I got a call from Ken one Friday afternoon. It was to the point but with caveats. 'Right, you are on to provide accommodation units at Bretby. I will be sending an architect to see you, who will design the conversion. But no one except you has to know the real purpose of the conversions. So, you must design a cover story. The real reason will come out later.'

The meeting with the architect took place the following week. He briefed me on the experience of his company and their intention to give the project top priority. He explained to me the percentage of the contract price that they usually took as commission. My experience at Selby sprang into focus. Surely, I explained, this was not like a normal building contract — there were no risks from ground conditions or environmental issues, these had all been resolved in the construction of the original building; this was a simple conversion contract. After some discussion he reduced his percentage claim by 2%.

The next issue to solve was the cover story. We decided to use Toyota and their interest in the facilities of the Demonstration Hall, as the cover. When there were staff from the architect's company on site measuring up in the Demonstration Hall and seeking plans and information about water and electrical resources, they were accompanied and supervised by a senior member of our team. That person made sure that any questions about the purposes of the visits and inspections raised by Bretby staff were told that it was a confidential survey and evaluation as a follow up about the possible future use by an international company.

The detailed plan was to create a series of apartments to four-star hotel standard, with up to date en-suite facilities which could be used for single occupancy or easily adapted for double occupancy when staff had their wives attending for a special final night event. I was not involved in the building works to provide the overnight accommodation, as it was decided to reduce the concentration of staff on the Bretby site and this would allow for the Staff College staff moving from Chalfont St Giles to Bretby.

The Research staff and Scientific staff remained at Bretby, but the rest of the Technical Department staff moved to Eastwood Hall (see below), which also had the field offices for John Northard and Ken Moses. This was very convenient for having face to face meetings with Ken Moses and, along with Colin Williams and Martin Shelton, we became closely involved with the main commercial and political issues facing British Coal. My stay at Eastwood Hall was less than a year, and during that time Colin Williams indicated that he would like to retire. Ken Moses suggested that I might like to take on the Supply and Contracts Department. We worked well together, and I think he preferred to have me rather than someone else imposed from within the corporation. He had an experienced mining engineer, Eddie Hindmarsh, who was then in the senior management team in Technical Department, who could take my place.

## Eastwood Hall

Eastwood Hall had been the Area office headquarters for the South Nottinghamshire Area of British Coal. Its location was just off the M1, at Junction 24A, and for me was a significantly shorter journey than to Bretby, as it removed the challenges of the traffic congestion around Derby. Obviously, for some staff who lived south of Bretby, around Burton-on-Trent, it presented a much longer journey than their travel to the Bretby site. The relocation presented the opportunity for some staff to take redundancy. I was delighted that Jenny Smith was able to transfer to Eastwood, so my support remained secure. Some other senior PAs also transferred, but we had to shuffle staff around to get the Technical

Department functioning effectively at the new location. There were continual changes taking place in the coalfields, as mines were closed, and it is difficult to reflect the fluid state which persisted throughout British Coal at that time. This was just before the initial proposals were tabled to consider privatising the coal industry.

When I left Technical Department, after nearly six years, the staffing levels were reduced by nearly 70% and the costs of running the Department were similarly reduced. At the same time, the organisation was modified, and the workload repackaged so that an effective service to the mines was maintained.

In 1990, Sir Robert Haslem, the Coal Board Chairman, had been awarded a peerage and elected to the House of Lords. I sent him a letter of congratulations from Enid and myself as we had met him and his wife on several occasions. He sent me a reply dated 26 June 1990, thanking me for our good wishes. Handwritten at the end of the letter, he added: 'You personally have done a marvellous job in reshaping and streamlining a wide range of our technical and development activities — cannot speak too highly of what you have achieved! Warmest Regards, Yours ever, Bob'.

Wow. Praise from the Boss!

# Chapter 18

*Head of Supply and Contracts Department
April 1991-November 1992*

*Overview*

My contract to become Head of Supply and Contracts Department started on 1 April 1991. The department was located in Coal House, a nine-floor building prominent in the centre of Doncaster. As far as I was concerned it was a local office, compared with the distant location of my previous offices at Bretby and Eastwood.

The Department had responsibility for buying centrally, to a national specification, those products in common usage throughout the corporation. It was also responsible for buying all mining machinery and spares. The department was responsible for letting all works and service contracts. The branches of the department had an annual spend of over £1.2 billion.

Also, the department ran six central stores for consumable items used throughout the industry.

Within the department was a section called the Plant Pool. This section managed a central holding unit which leased mining plant to the pits and other operations. The Plant Pool was responsible for accepting used plant back from the collieries and arranging, as and when required, for its repair back to new standard.

Most of the repair work was undertaken by the six central workshops, which were located adjacent to the central stores. These workshops formed a further branch of the department, with an annual turnover of £95 million and they were accredited to BS5750 standard. When I arrived, the workshops were beginning to undertake contracts with clients outside British Coal, particularly among the oil and gas operations in the North Sea.

My initial impression was that the staff of the department operated to the highest business standards and were proud of their skill and experience at negotiating sound commercial contracts for goods and services for the industry. I also inherited a very able private secretary, Diane Izzard, who was very experienced in the range of operations taking place throughout the department. She also proved very proficient in supporting me with my other external commitments, particularly as President of the Institution of Mining Engineers. In my time in charge, I never encountered anything untoward to make me change that view. The senior staff heading the different sections of the department formed a cohesive team that worked together to tackle the challenges facing British Coal as they impacted on the Supply and Contracts Department.

### Opportunities for change

Just like Technical Department, though, it was clear that the contraction of the coal industry required that the Supply and Contracts Department needed to slim down its staff to reduce its costs and this was accepted by the senior team.

There were some clear options. The department had its own printing section which produced all the documents that were used throughout the industry to initiate and monitor requisitions for plant and stores. The volume of documents required was obviously reducing. The answer was to privatise that section and hive off its staff and their equipment. This was done, with safeguards for the cost of the documents produced for British Coal, but the team were able to take on printing work for outside bodies and companies. As the operation of the new printing plant was established on a site away from Coal House, it formed an initial vacancy in the building that indicated that a relocation of the whole department away from Coal House could be a major cost reducing opportunity (more of this later).

## Steel; a purchasing challenge

British Coal was a major user of steel for roadway supports and other structures. At an annual review, shortly after I arrived, the negotiations with British Steel were deadlocked as they were demanding significant price increases way beyond the general levels of increase being applied in other businesses. We had numerous discussions about this issue within the senior team. We had to find an alternative supplier at a lower price. After some secret investigations, a company in Ireland was prepared to sell us steel of the right quality at a satisfactory price. I had briefed Ken Moses on the issue as it developed, and I told him that we were going to place orders with the Irish company, and he agreed. The tactic worked, and after a short time British Steel came back to the negotiating table with an acceptable price offer for their steel products.

## A reducing business

It is difficult to recognise the reduction across the board of the need for consumables in the coal industry at that time. As pits were closed, the demand reduced naturally with less collieries in operation. But in some parts of the country the salvage of good quality, little-used equipment, as an example conveyor belting, could be moved to continuing pits at a significant cost saving to the operating colliery. This meant that our purchasing levels of some consumables were reduced to a much lower level than formerly. It also resulted in the number of suppliers being reduced. Where we normally liked to have several suppliers, so that there was competition, we were able to operate with one or more less. It was a continual challenge to review the stock levels for all items held in the central stores to ensure that we carried the appropriate stock levels to satisfy the needs of the pits still in production.

It was understandable that we would receive visits by the senior management of our main suppliers trying to establish what their role was to be in the future of the coal industry. Some of the major companies had significant business overseas, where they could expand, but others had no option but to slim down their manpower and try to get work in other UK industries. Some had no option but to shut down. It is a responsible role for staff dealing with suppliers to give them truthful information about the trends in the industry and how those trends might affect their own company. With the coal industry contracting quite quickly, the majority of the information would be sombre news, although there were odd instances of good news, where one supplier was being trusted to take on the total supply of their product at a competitive price.

*Central Workshops*

Mining is a tough environment underground, and the equipment is distributed over many miles of roadways. The underground equipment has two roles: it has to transport men and materials to the point of production; and it also has to transport the coal, which is a bulky product, to the pit bottom so that it can go via the shafts up to the surface. By the late-1980s the application of heavy duty coal faces was establishing rates of production on some coal faces which meant that conveyors had to be significantly upgraded, and underground bunkers installed to compensate for any stoppages with the winding operations of the coal shafts.

The size of this change is best understood by a bit of history. In the 1950s, when I started in the mining industry, the rate of production was measured by how much coal a face worker could shovel in an hour. That figure was about three tons. There might be 20 men on a coal face, so the rate of production from the coal face was around 60 tons per hour. The new heavy-duty shearers, with ratings of 400 horsepower, could produce coal at the rate of 500 tons per hour on some of the thicker coal seams being worked.

All the equipment throughout a mine is worked hard, and it is inevitable that some machines will fail. Changing a machine, or a section of a machine, on a coal face is difficult and challenging by the very nature of making the roof conditions safe for the people involved to work. Repairing machines on thin seam coal faces requires midget craftsmen and the ingenuity of brain surgeons to fit the pieces together. It is logical that the aim should be to use reliable equipment and limit breakdowns to a minimum. Over the years, the central workshops had established standards of repair to the various machines put to them for repair, such that they left the workshops as good as a new machine from the manufacturer. Elaborate tests were carried out under load to prove that that standard had been achieved.

Each of the six workshops had developed specialist expertise on the equipment they repaired. Each workshop covered different pieces of equipment — shearers and coal face equipment; development machines for driving roadways; conveyors; electrical equipment; ventilation equipment; electrical cables and control boxes. In every case, the particular workshop monitored the units they repaired and sought to increase the reliability of their products.

There was a proposal to adopt planned maintenance by making changes to machines on a routine basis to avoid breakdowns. How to agree an appropriate plan, however, was very difficult. Different factors were considered: time in use; tons of coal cut and tests on the unit to establish wear on key components. It was not applied widely in the industry.

However, at the Tursdale workshop, in Durham, they had developed an alternative approach. On some electric motors, they carried out vibration tests when the motor was repaired. They recorded these results and printed them out. They then designed a small vibration monitor which could be taken underground and applied to the electric motor and its results retained. If those results were the same as the vibration tests when the motor was newly repaired and tested in the workshop, then it indicated that the motor was healthy, and it could remain in use. There were other initiatives in the five central workshops to improve the reliability of the plant in use. With the development of micro-electronics, the possibility for measuring the health of machines was starting on an upward trend which the central workshops and their staff were well equipped to utilise.

Visiting the different workshops and talking to the management team and the workmen was a routine part of my role, and I was most impressed by what I saw and learned of their achievements.

How were these pockets of expertise in the different workshops to be preserved for the future of a smaller coal industry? That was clearly a question for me and my senior staff to answer.

## Voice problems

When I moved to Coal House in Doncaster, I had developed voice problems such that, by the end of each day, my voice was failing, making discussions in meetings and telephone conversations quite difficult. At that time, I was Vice President of the Institution of Mining Engineers and was scheduled to become President in January 1992. As President, I could expect a year of special events, with numerous public speeches to make, in addition to my day job. I began to wonder if I might have to abandon being President. I discussed this with my son, Brynnen, who was a GP. His reaction was to seek a solution to the voice problem, rather than consider not taking up the Presidency. He arranged for me to attend the speech therapy unit at Doncaster Royal Infirmary, which was situated very near Coal House.

For a few months, I attended weekly sessions held by a speech therapist. She did an excellent job training me how to breath and speak without forcing my voice. It was only later that I realised that the cause of my vocal problems resulted from excessive exposure to the effects of air conditioning. At Bretby, I'd had an air-conditioned office and I also had a car with air conditioning. It was the same with my car and my office in Coal House. Subsequent to this period, I always tried to restrict my exposure to air conditioning.

## Plant Pool

The Plant Pool organisation was a major unit within the Supply and Contracts department of British Coal. Its objective was to utilise the resources of plant and machinery to maximise efficiency and reduce costs. Plant was leased out to the users at the pits and other operations at a range of charges to reflect the size and costs of the different pieces of plant. When the plant broke down or ceased to be needed it was required to be returned to the Plant Pool as quickly as possible. Only when the item was returned to the Pool did the hire charges cease.

In my time, the Plant Pool was managed by Ron Pine. Ron had been a Production Manager in the Selby Coalfield when I was Director there. He was a very responsible mining engineer who understood the changes taking place in the coal industry, and he had to use all his skills to keep the size of the total Plant Pool appropriate to the size of the industry. He also had to be able to supply spare units when there were breakdowns or there could be major shortfalls of output. He also had to liaise with the workshops managers to have an appropriate flow of repaired units back into the Pool to satisfy the demands from the users for new installations as well as breakdowns. He also needed to be aware of new initiatives and trials of up-rated units which could make items in the Pool obsolete. Ron and his staff could be described as a set of super jugglers with lots of balls (units of plant) up in air all the time, not quite sure when the next call would come for which piece of plant!

### Seeking increased efficiency

There was a similar requirement in the Supply and Contracts Department, as applied in the Technical Department, namely to make changes to increase efficiency and reduce costs. There was a difference in the make-up of the Supply and Contracts organisation, though, in that it had many more different sections. As the costs were primarily wages and salaries, there was the need to reduce the headcount. It was appropriate to have a plan to 'nibble', wherever possible, in the many sections, to release a few people. The senior management team was committed to this procedure, and at the same time they made structural changes to streamline and simplify the organisation. The redundancy terms available then in the coal industry were attractive, to allow willing people to gain a lump sum and clear their family debts and make a fresh start in a different job. It had to be a continual process and there was a long way to go to keep pace with the reduction of the number of mines.

## British Coal as a landowner

British Coal was a major land owner in the UK. Its need for land at the operating collieries was extensive due to the demands at the mine for land for forming tips for the waste material from the washing process, along with land for lagoons to clean the waste water. There was also a need for land for stocking the materials used underground and on the surface of the mine. Due to flexibility of demand for the different qualities of coal, many of the large mines also needed stocking grounds for run of mine coal or the different washed products. So, the closure of any colliery potentially released large areas of land.

Opencast mining was still operating as a successful and profitable producer of coal, and sites were managed in British Coal by the Opencast Executive. These sites were in areas of the country where the coal seams were near the surface in what were designated as exposed coalfields. New equipment and operating methods in opencast had allowed the coal to be extracted to greater depths. There were significant areas of the country where opencast mining could take place. In some cases, the land on the surface was owned by British Coal, but there was a tradition, in some parts of the country, to operate sites where the land was taken over from the farmers during the opencast operations. The land was retained for a further five years after the coal was extracted, to restore the land for farming. The whole site was initially provided with a competent drainage system. It was also cleared of stones and other trash and farmed with variable arable crops. During the five years the land was enriched with fertilisers and soil inputs such that it could be returned to the farmers, in many cases, in better condition than it was prior to the opencast operations. It was specialist contractors who carried out the work during the restoration period, under direction of agricultural experts.

One other technology had been developed which affected the surface of the mines. In many cases it was known that there was a percentage of coal in the pit tips. This could be evaluated and could be as much as 5% to 8%. A simple form of barrel washer was used to extract the coal. The washer was effectively a rotating cylindrical drum into which the raw material from the tip was loaded and the coal, which was floated out, was separated from the dirt. A site could be worked by a small team of men and the saleable coal on an efficient site could give substantial profits to the operator and British Coal. These operations provided the option to take the residual dirt from the barrel washer and re-profile it into an entirely different format compared to the initial tip.

In essence, there was operating technology available to completely remove any signs of coal mines and make available the colliery sites for developments like housing and other operations. In the exposed coalfields the formula was: plug the shafts, remove the surface buildings, opencast the site and mine all the coal in the shaft pillars, barrel wash the tip and re-profile the site and sell it for a new use. In many parts of the country there are now no signs of the old mines. This is very different from the position of former coal mining areas in other countries. In France there are cases of old coal mines just being abandoned, and their sites are still on display as an ugly blot on the landscape.

By contrast, in the UK, there are examples of sites which are remarkably converted. The area in South Yorkshire encompassing the Manvers and Wath collieries, along with the other associated operations, described above when I was in South Yorkshire Area, are a classic example of change. They are now an expanse of new roads and an array of large, impressive modern buildings housing new businesses. Similarly, in the recovered area of land between Grimethorpe and Houghton Main Collieries, is the massive central distribution operation of ASOS. This international company sells all its clothes products around the world, from that one site, through internet trading.

One of my early jobs in Supply and Contracts Department was to attend a meeting in South Midlands Group to rule on a dispute between the Group and a Barrel Washing Contract operated by Ogden mining. The meeting was attended by Robert Ogden himself, whose company was very successful in other barrel washing operations. However, in this case the relationships had soured beyond repair and I had to rule against him. At the end, Robert said it was sad that this problem had not been exposed earlier, when it could have been resolved without the contract being terminated. I agreed with him, but I didn't express my view openly in the meeting. It made me wonder if the Group Director, Tony Deakin, who had followed me as DDM in South Yorkshire, saw this as a test case for me in my new post. Robert Ogden was noted as a very significant figure among the private companies doing business with British Coal. Tony might have doubted my ability to rule in his favour against such a character as Robert Ogden.

## The challenge of 1992

As 1992 approached, I realised it was going to be an important year in my life. I wrote to Ken Moses and asked him to obtain permission from the Board that I could take up the responsibilities as President of the Institution of Mining Engineers, along with my job as Head of Supply and Contracts Department. I got approval, and in many ways the Board gave me continued help through my year in office.

Being President incorporated chairing the Council of the Institution and some of its key committees. It also required meetings with the Secretary of the Institution on business matters and the annual programme of meetings and events.

The President was invited to speak at the annual dinners or dinner/dances at the local branches of the Institution. There were six branches — Scotland, Wales, Staffs and South Midlands, Notts and North Derby, Midland Institute (Yorkshire), and the North East. There was also a Junior Section of Notts and North Derbyshire.

The major commitment for the year was to the Annual Conference for the Junior section, followed by two days for the Annual Conference for senior members. Both of these had technical papers, and special evening events. There were a range of special events for the ladies attending both conferences.

The whole week culminated with the Annual Banquet and Ball.

Immediately in the New Year I had to get involved in the Institution commitments. Within the Supply and Contracts Department, among the senior staff, there was a genuine interest in this prestigious programme, as the department had not seen anything quite like it before. Two or three of the staff were keen to have input into the speeches. We decided to try to do some research to provide new ideas. The theme of the Presidential address, 'Preserving the Heritage' focussed on the historical aspects of the mining industry. We knew that the industry in the 1990s was very different from the industry before the First World War and we set out to identify the total output from the South Wales Coalfield and from the Durham and Northumberland coalfields, from their start to the present time. The findings were a surprise for all of us and were key facts for both the speeches.

Once the diary was fixed it was clear that I was being given a major test, as I had to propose the toast in Wales on Thursday 19 March, at the Midland Institute in Sheffield on Friday 20 March and in Scotland on Saturday 21 March. If the programmers were tipping that I would use virtually the same speech on the three nights, we had to make sure they were proved wrong. All the speeches would be different.

The arrangements for the Conference programme were not straightforward. We had, in the past, started a Conference with a church service. When Bill Forrest was President, the service was in Coventry Cathedral. I suggested a church service to John Bourne, Secretary of the Institution, and he rejected it as, he said, there was no Church near the

Pembroke Hotel which was the centre for the senior conference. I travelled to Blackpool with Gwynne Richardson, Chaplain for the Institution, and we found that St Paul's Church was within a hundred metres of the hotel. It was also an ideal size for the sort of congregation we expected. Gwynne arranged the church.

Throughout the preparation and execution of the Annual Conference, the President of the North Staffordshire Institution, Frank Middleton, and his colleagues, worked ceaselessly to ensure it was a memorable success. For the opening church service, they obtained the support of the Point of Ayr brass band and the Parkside Colliery Male Voice choir. This ensured that the music and the singing were of top quality. The address was given by Right Reverend Alan Chesters, Bishop of Blackburn, who'd undertaken research to make sure he understood the role of the Institution of Mining Engineers and the uncertain future for the coal industry in 1992.

The theme for the Conference was 'Assets for Any Future'. I was able to get senior people within British Coal, and colleagues who could give excellent technical papers on that theme, which would attract senior staff in the industry and in businesses that served the industry.

The total programme was as follows:

'Preserving the Heritage' Presidential Address — C T Massey.

'The Culture of the Future' by Mr J N Clark, Chairman of British Coal.

'The Technology for the Future' by J C H Longden.

'The People for the Future' by Mr K Hunt.

'Closing Address' by Dr K Moses, CBE.

Due to the fact that several members of the Board were involved in key roles at the Conference, the Chairman, Mr J N Clarke, decided that the whole Board would move to Blackpool so that they could carry out their business responsibilities as well joining in the Conference activities. This decision gave the Conference a boost and attracted support from our main suppliers as they welcomed the opportunity to see and be seen by Board members. It suggested that this conference might be a special occasion.

## Three speeches in three days!

This was a logistical, as well as a speaking, challenge. Wales on Thursday 19 March, Sheffield on Friday 20 March and finally Scotland (Edinburgh) on Saturday 21 March — a chance to have three speeches essentially the same, or, if they were different, making sure I had the right text on each night.

This intensive three days were a success. There were two other Branch speeches to make before the week of the annual conference — Notts. & North Derbyshire Branch on 3 April, and South Staffs & South Midland Branch on 24 April.

A résumé of the speeches is included as an Appendix to this book.

The completion of the speech to South Staffs & South Midlands branch gave a few days of rest and anticipation before the Annual Conference. But there was a surprise.

## Letter from the Prime Minister's office

The letter was dated 8 May, 1992.

> *'Sir,*
>
> *The Prime Minister has asked me to inform you, in strict confidence, that he has in mind, on the occasion of the forthcoming Birthday Honours, to submit your name to The Queen with a recommendation that Her Majesty may be graciously pleased to approve that you be appointed an Officer of the Order of the British Empire.*
>
> *Before doing so, the Prime Minister would be glad to be assured that this would be agreeable to you. I should be grateful if you would let me know by completing the enclosed form, and sending it by return of post.*
>
> *If you agree that your name should go forward, and The Queen accepts the Prime Minister's recommendation, the*

*announcement will be made in the Birthday Honours List. You will receive no further communication before the list is published.*

   *I am, Sir,*
   *Your Obedient Servant*
   *Xxxx'*

Wow! What a surprise, but of course it could not be shared with anyone except Enid. There was also the uncertainty that the Queen might not accept the Prime Minister's recommendation!

## Relocating the department from Coal House

An appraisal of the costs of running the Purchasing & Supplies Department clearly showed that a significant saving could be made if the headquarters could be moved to another Coal Board site in order that Coal House could be returned to the local authority. There were many Coal Board sites becoming available as a result of the colliery closure programme, however most of these sites were in isolated locations and had only limited office buildings. There was no intention that new buildings would be erected to provide new offices for all the department staff. It was also considered that the new location should be in Yorkshire, so that it was not too much of a distance from the centre of Doncaster for current staff to travel.

In the end, it was agreed that the new location would be at the Fence site, which was just off Exit 31 of the M1, about half a mile towards Sheffield. It was given the grandiose name of The National Provisioning Centre at Fence. The site also included a central stores and a central workshop. It was a major job to house all the different sections of the department and establish the communication links that allowed business to be done with all sites throughout British Coal and to all our suppliers and manufacturers. Excellent work was done by our senior team to achieve this objective and the National Provisioning Centre became operational on 10 August 1992.

The changeover resulted in significant staff savings, with numerous people opting to accept the redundancy terms which were available. I was fortunate that my secretary, Diane Izzard, moved over to Fence and she was able to continue to give me her excellent support right through to my retirement.

### The Annual Conference of the Institution, Blackpool, 11–15 May, 1992

The Institution of Mining Engineers Junior Sections' annual summer meeting took place on the Monday and Tuesday of the week, with technical papers and ladies events. I made an appearance at the Annual Dinner on the Monday night, and I enjoyed meeting the junior members and talking to them to try to allay some of their fears for the future of the coal industry.

Wednesday was a quieter day, with the Commodore King Challenge Cup golf competition and the Civic Reception in the evening. It was reassuring during the evening to welcome my eldest son and his wife and my daughter and her husband, who were there to give me morale support and some active help during the next two days.

The Service of Thanksgiving was the start of the Conference on the Thursday morning. We saw the church was comfortably full as we waited to enter, while the Point of Ayr Brass Band completed their pre-service programme. They then played a fanfare, the congregation stood while the Presidential Party, the Bishop of Blackburn, the Vicar of St Paul's, and the Chaplain of the Institution paraded to the front of the Church.

The Presidential Party was:
The President and his Wife.
Mr L J Mills and Mrs Mills — Honorary Treasurer of the Institution
Mr F Middleton and Mrs Middleton — President of North Staffs Branch.
Mr W J W Bourne, OBE — Secretary of the Institution.
After the welcome, the service moved forward following the programme. In the first hymn the choir alone sang the verse:
*Our midnight is they smile withdrawn*
*Our noontide is thy gracious dawn,*
*Our rainbow arch thy mercy's sign;*

*All, save the clouds of sin are thine.*

After the prayers I read the first lesson.

The band then played the hymn tune 'Gresford', which was composed to honour the miners killed at the Gresford colliery disaster. It is a quiet, moving piece of music.

Gresford mine was in Cheshire, not too far from Blackpool.

Frank Middleton read the second lesson, which was followed by the hymn 'Love divine, all loves excelling', to the Welsh tune Blaenwern. The singing, by the mainly male congregation, and the male voice choir sounded truly triumphant in the church. There was a feeling of expectation starting to arise in the congregation that the service was becoming an appropriate way to start the conference.

The address by the Bishop of Blackburn did not disappoint. He had done his research, and his words struck the right balance of respect for the past and hope for the future. It was followed by the male voice choir singing 'Morte Christy', written by Emrys Jones. This Welsh song carried forward the triumphant sequence.

The prayers of intercession, led by Gwynne Richardson, the Institution Chaplain, included responses by the congregation. The final congregational response was:

*'Eternal God and Father,*

*You create us by your power and redeem us by your love;*

*Guide and strengthen us by your Spirit*

*that we may give ourselves in love and service to one another and to you;*

*Through Jesus Christ our Lord. Amen.'*

In the final hymn, 'All people that on earth do dwell', again one verse was sung by the choir. This time with a show of power and triumph.

*O enter then his gates with praise;*

*Approach with joy his courts unto;*

*Praise, laud, and bless His name always,*

*For it is seemly so to do.*

After the blessing by the Bishop of Blackburn the choir sang a vesper,

'Spirit of the Living God'. They sang it pianissimo, and a spirit of peace and calm descended on the church.

It was not the end, as the choir then sang 'Gwahoddiad', a Welsh song of power and praise.

The final item by the band was 'Procession to the Minster' by Wagner. This piece starts very quietly and then it slowly gets louder and louder. There are two crescendos near the end when the euphoniums and the horns rise up through the band playing double forte. I was watching the conductor as they played this item. He was leaning over, looking at the band, and revelling in the sound they were making in the church. He knew that they were on fire and playing as good as they had ever played.

To the sound of a fanfare, the ministers and the Presidential party moved off to the south door of the church. There were favourable comments from many members of the congregation, but I will always recall the comments of a Scottish company representative, 'Did you initiate this service? Were you behind the format of it?' he asked me. I confirmed that I had been involved and I was pleased that we had required the service to be a part of the Conference. He shook my hand and said, 'It was a moving triumph in every way. I will never forget it.'

The next challenge was to present the Presidential Address, 'Preserving the Heritage.' There was a good attendance, and John Bourne asked the audience to rise to welcome the President. John Bourne and myself walked past the audience to the stage. I was accompanied by my son, who was a stranger to the audience.

I presented the introduction to the address.

'Being President of the Institution carries with it the obligation to present a Presidential Address. This is the eightieth such address (not every President has given one) and some members could well be forgiven for groaning and saying, 'not another Presidential Address!'

The Address gives the author an opportunity to focus attention on some subject or theme, important at that point in time against the background of major historical events affecting the country. There is no consistency with this.

In 1918, G W Walker referred to the Great War and said, 'we have, therefore, great cause for thankfulness and pride for the part which the mining communities have borne in this war.' Similarly, in 1945, Professor Douglas Hay rejoiced at the end of the Second World War and captured the uncertainty facing the industry pre-nationalisation.

Neither wars were anticipated in Presidential Addresses. Pre-First World War President W E Gosforth gave a technical paper recording the history of the experimental work in a gallery at Altofts Colliery in Yorkshire, on coal dust explosions and the need for stone dust to dilute the coal dust. In 1939, E O Foster-Brown gave a masterly review on the coal question, comparing the performances in Great Britain with continental producers and the United States, but he does not anticipate a European war.

'Gentlemen, I have elected to try to formulate an address relevant to this point in time, but reflecting an historical perspective. My reasoning results from a concern that we are all becoming dominated by today's problems. In fact, the present is short and transitory; the past is long in time and factual; the future may be uncertain, but it is also long in time. There is a great risk, always, that an instant solution to today's problems will severely limit the options for the future. The options for the future – any future – need key resources. The industry has resources in abundance, but can develop them further. These key resources will be the subject of presentations later at this conference, and I am indebted to the authors for their support.

The future must evolve from the past, as well as the present. We must not forget the heritage which we have received from our forefathers, and we must ensure that we pass it on. Hence the title of this address, 'Preserving the Heritage'. A heritage is defined as 'an asset that passes from an ancestor to a descendent.' It is only possible to discuss a few of these assets but I hope to cover sufficient to demonstrate that the heritage of the coal industry is of great value.'

I moved to the side and my son Brynnen read the main body of the address. The subjects covered were 'pure Science', including details of explosives used in the industry, 'Engineering experience', included the development of steam power in the northeast of England to solve pumping problems, the adaptation of steam power to drive the railways, and the conversion of steam to generate electrical power. The application of electricity in the mines had led to the shearers in the 1990s being able to produce coal at the rate of over 500 tons per hour. The engineering section also noted the development of the powered supports which were evolved from the hydraulic undercarriage of the Wellington Bomber.

There was a section on 'Safety and Health', which showed the reduction in fatalities in the industry, but it also described the technology being used to overcome the hazard of dust on the coalfaces.

The 'business' was then described. It showed the size of the assets taken over at nationalisation — nearly 1,500 mines, 30 briquetting plants, 55 coke ovens, 85 brickworks, 1,803 farms, 140,000 houses, 27000 farm houses and agricultural cottages, 2,785 shops and business premises, a cinema and a slaughterhouse. There were also private railways, wharves, coal depots, milk rounds, a holiday camp and a cycle track.

There was a section which listed the output over the years, and further charts showing the change in the markets for coal, including exports.

The final charts showed the effect of alternative fuels replacing coal in the marketplace. The result was the downsizing of the industry. Between 1957 and 1973, output at collieries was reduced by 37%, colliery manpower reduced by 62% and the number of collieries reduced by 65%.

Bryn then moved over and I covered the conclusions.

'In an address of this length, there can be criticism for selectivity in the examples used. The justification for the subject matter in this address, and the examples quoted, derive from the desire to counter the public image of the industry. The industry was so big, at the start of the twentieth century, that it employed 10% of the total working population in the country, and it was therefore bound to have a high profile. Its disasters, public enquiries, industrial disputes and bitterness, spilled over from the coal industry to affect the whole country. However in the 1970s and 1980s, the ordinary individuals have some justification for wishing for a change. One can imagine them saying 'have we always to put up with this industry? — it is an old and worn out industry anyway.'

Gentlemen, it is not worn out; it may be old, but that age reflects a sound professionalism within the industry.

In this address I have tried to illustrate the scientific base of the industry, its engineering skills, which are at the forefront of technical standards, its safety and health record, which gives a lead to international standards, and finally its experience at facing up to change in the commercial challenges. It is a wonderful heritage, of which all mining engineers can be justly proud; it needs preserving intact and then passing on to the next generation.'

I was satisfied that the address had been well received by the audience and that it had certainly held their attention. What I did not know at that time were the questions being asked about the young man who had read the main body of the paper. Who was he? Which department of British Coal did he work in? They were impressed by way he had presented the arguments and the illustrations. I knew Bryn had spent some time practising, so that he understood how to get the meaning over. I was able to thank Bryn in my comments at the final event of the conference.

The rest of the morning and afternoon was taken up with two technical papers. I chaired the first one, which was given by Mr J N Clarke, Chairman of British Coal, entitled, 'The Culture for the Future.' My friend and colleague John Longdon gave the second paper, 'The Technology for the Future.'

After the formalities of the day, the evening was listed as an informal affair, with entertainment ('Something Special') and dancing. The something special was a comedian, Tom O'Connor. He was brilliant, and he charmed the audience with lots of stories about life and some of his time as a schoolteacher. He rounded off what had been for me a very successful day.

On the Friday, there was the final paper by Mr K Hunt on 'The People for the Future.'

This was followed by the closing address by Ken Moses. It was in Ken's typical style — new ideas and challenging comments interlaced with some humour. It was shorter than I expected, but it gave a brief gap up to lunch and then an afternoon at leisure before the Annual Banquet and Ball.

Murdo Spence had not been able to attend the conference, but I had asked him to propose the toast to the Institution. It was a relief when I knew that he'd arrived.

Murdo's speech was full of humour and he had a few digs at me in my Coal Board role, which the audience loved. My job was to follow Murdo, and propose the toast to the guests. I knew that Murdo would bowl me a few googlies.

'Madam Mayor, distinguished guests, ladies and gentlemen, and Murdo. I should have known when I invited Murdo to propose the toast to the Institution that I would stand here in absolute isolation, and all of you are thinking 'follow that!'

As is usual on these occasions, Murdo had appeared in his dress kilt. In Supply and Contracts Department there was always some concern when Murdo appeared in his kilt. Most people appear at negotiations with something tucked up their sleeve, but Murdo has much larger margins tucked up his kilt.

'We welcome Murdo and his wife, Maureen, here tonight. Murdo has for a long time been one of the great characters of the mining industry. His skill as a mechanical engineer, his wit, his enthusiasm and his idiosyncrasies have been a memorable part of this industry for many years. I think that we all admire how Maureen is able to cope with the Murdo idiosyncrasies. Murdo, I do thank you for proposing the toast to the Institution.

Ladies and gentlemen, my decision this year to seek the speakers for the technical papers at the conference from the most senior people within British Coal, was a total success. I am deeply thankful to John Longdon and Kevin Hunt for giving the technical papers, and for Dr Ken Moses for giving the closing address. I am particularly honoured that the Chairman of British Coal, Mr Neil Clark, gave a prestigious lecture following the Presidential Address.

The Conference has benefited from a host of other people. The North Staffs Branch and its Secretary have worked for months to ensure that this conference proceeded smoothly. Their support, along with the permanent staff of the Institution, has been invaluable. There has been support from the other parties in the great mining family — the manufacturers, the suppliers and the mining contractors. Their input has been an immense help and is most gratefully appreciated.

We are delighted that the mayor and Mayoress of Blackpool have been able to join us tonight. We are very appreciative of the reception which they hosted earlier this week.'

The speech then listed the kindred institutions and associations present.

'I am also happy tonight to have two of my children here, with their partners. They were brought up, in their early days, to be involved in the operations of the mining industry. It was not considered strange, therefore, when the idea was put forward that Brynnen should read the major part of the Presidential address, to spare my failing vocal chords. I thank him very much for his help.

My final guest, who has yet to sing for his supper, is Tony Hewitt. We were born in adjacent villages in Yorkshire, and we both went to the same school in Wakefield. We worked together as colleagues in the late 70s in the South Yorkshire Area. He later had outstanding success as Chief Executive of British Coal Enterprise. His acclaim outside the industry led to him dining in the House of Commons and also being entertained in the House of Lords. These experiences were nothing compared to him being entertained to dinner in his local town of Barnsley. This made the front page of the Barnsley Chronicle, which is usually reserved for the famous sons of Barnsley, like Geoff Boycott or Michael Parkinson. So, Tony, we welcome you here tonight, with your wife Lucy, in the knowledge that, if you have made the front page of the Barnsley Chronicle you have really made it now, mate!

Finally, I have to welcome the ladies. Resplendent as ever, they grace our functions and give style to the whole Conference. But more important, all of us — the industry and the Institution — owe them a debt for their companionship and support which is priceless.

Members of the Institution, would you please charge your glasses and be upstanding as I propose the toast to our guests.'

It was a relief to get the end of the Annual Conference without any major mishaps. Back to normal, there was a new issue going live.

## Privatisation of the coal industry

The rumours about the possible privatisation of the coal industry had become a reality, and at Headquarters, a senior staff team was deployed to prepare position papers. One part of the Supply and Contracts organisation that might possibly be able to stand alone and operate outside a privatised coal industry was the central workshops. They all had experienced workers who were true craftsmen, the workshops were well equipped with machinery and modern techniques for testing, which guaranteed the quality of their repair work. Also, some of them were already doing work for companies outside British Coal. Might it be possible to sell the workshops off to a private organisation or company before the full privatisation of the coal industry?

The problem was that, in the field, we had no idea what form was envisioned for the planned privatisation, or its timing. Clearly, the key decision maker on these questions would be the government. However, there was interest by individuals and companies in the workshops as going concerns. As the year progressed, some parties saw the possibilities and their interest became quite serious. They visited the sites and witnessed the operations and saw the quality of the repaired units sent out for re-use in the Coal Board and other industries. I encouraged this interest, and I spoke to the main contenders. I was motivated by the wish to keep the skilled teams together, rather than seeing them broken up, with the years of hard work to foster and evolve top class engineering workmanship completely destroyed. My personal involvement gave confidence, for a period, to the potential buyers. It also gave hope to the workshops' staff that they might have a future in their current jobs. It was not possible to get any clear guidance from the Board on a future policy for the workshops and, when it became known that I was to retire, the interest of the buyers and the hopes of the workers died.

## A delicate family challenge

It was natural that Enid and myself wished to have the family around us when the Honours List was announced on the Queen's official birthday. Our three children initially had plans and commitments for that day. There thus began numerous phone calls by Enid involving subterfuge about the false reasons for a family get-together, until they were all brought into line. When Adrian arrived from Stockport, he said, pointing to the lists in his paper, 'Now I know what all the fuss was about.'

We had a family celebration on the day of the announcements and it was followed by much publicity. The award of the OBE was a very great honour and I was humbled by the many letters of congratulation that I received from within British Coal and companies associated with the coal industry. There were contacts from many people I had known in the local community and at universities, and even in government departments.

It was clear that the invitation for Enid to go to the Investiture would only allow for two of our children to accompany her. The three of our children immediately agreed that Enid should be accompanied by my mother, and none of them would go. My mother, who was 85 years old, had attended the British Empire Exhibition in June 1924, at Wembley, with 700 workmates on a trip organised by the shirt factory where she worked in Barnsley. This would only be her second visit to the capital and she would be in the presence of the Queen!

## Another job?

Ken Moses invited me and Enid to have dinner with his wife, Mary, and himself. He gave me no indication why he wanted to see us. He had been very kind in his comments on my OBE. He said the award was overdue, which was an interesting reflection. The dinner was held at a hotel that Ken used for such meetings, where he was very well treated. The food was excellent, and we had plenty to talk about, with our families and the Institution of Mining Engineers, without discussing operations in British Coal.

Then Ken dropped a bit of a bombshell — he would like me to consider moving across and taking over the Coal Products Division of British Coal. This was a separate organisation, set up on nationalisation and covering the coking plants and pulverised fuel plants. It included processing plants converting the by-products into different grades of fuels for motor vehicles.

I knew little about the Division and had always thought of it as an offshoot of the British Coal Marketing Department. I rather gathered that Ken thought that I had successfully managed the reorganisation and downsizing of both the Technical Department and the Supply and Contracts Department and he must have concluded that Coal Products needed an outsider to move in and take an objective view of how it must be changed to be relevant for the future. It needed that person to be someone who would then manage the organisation to carry out the changes.

I was not sure that I wanted another new challenge. I had already decided that I did not want to continue into a privatised coal industry, whatever form that might take. So, it's fair to say that my reaction was cool, at best. There was some discussion, but Enid had made her mind up that she did not fancy me taking on a new challenge. She stood up and said, in a strange tone of voice for her, 'If you will excuse me, I am going to the ladies toilet.'

Ken got the message.

'That's it, forget it; consider that this suggestion was never made,' he said.

This was typical of Ken Moses; if he realised that his plan would not work he was always prepared to shut the door and try again. He did not hold it against me and we worked together as if the meeting had never taken place. But it was only for a short few weeks.

### Retirement prospects?

Within the corporation there were opportunities for senior staff to accept favourable terms for redundancy and early retirement. I had the full number of years in the pension scheme to qualify for the maximum size of pension available. I spoke to the person in Staff Department handling senior staff cases, Colin Ambler, who had been my Staff Manager at Bretby.

I expressed my possible interest in early retirement. The open question was to agree the best date for me to leave.

When I started the first discussions in September, we were still working hard in the new offices at Fence to make sure that the Department was fully functional in its new home. We needed another month or two to complete the satisfactory commissioning, which I wanted to supervise.

The retirement date needed to be near the end of my year as President of the Institution, which was January 1993, when I handed over to Albert Tuke as President to succeed me. Another factor which came into the reckoning was when the Investiture would take place for my OBE. John Northard gave me advice on this. The Honours were presented in alphabetical order and, in his experience, I could expect to be called to the Palace around November time.

Eventually, we settled on the end of November, which was preferred to the congested period for celebrations at the end of December.

### A major shock for the coal industry

On the morning of 3 October 1992, I received a call in my office from John Northard. It was unusual for John Northard ever to ring me, and this was the first call of the day. He ensured that I was sitting down and then he gave me the news. Ken Moses had collapsed in his office the previous afternoon, and he had died, aged 61. It was a shock for me and for all his contacts in the industry. What a tragedy for his life to be cut so short.

I knew that Ken had started to make preparations for his retirement. He had constructed a library and reading room over his garage, and no doubt he had discussed with Mary plans for travelling and other ways to spend their time in retirement. But I had realised that he was reluctant to make the change, he had so much to lose. He had pushed himself from the very first day he started work to get qualified and then get promoted. He was always sad that he didn't go to university, because he knew that he was capable of getting a good degree. Late in life he undertook studies and did a research project on the Selby Coalfield at Nottingham University. From these efforts he gained degrees of MPhil and a PhD, so achieving his lifetime ambition of becoming a university graduate.

In an obituary of Ken, he was described as a man of energy, judgement and penetrating intelligence. He had no time for muddle-headedness, but he also had compassion and great humour. He was equally at home making a technical proposal between a robust panel of his peers, or as an amusing after dinner speaker.

I would agree with this image of Ken; but I would add that he was a tough boss to work for, but he was fair and appreciative when you delivered the goods.

His funeral was taken in a humanist format which was a challenge for those who spoke about Ken. The large number of people who attended will probably take away the memory of the final item, Louis Armstrong singing 'What a Wonderful World,' — Ken's composite view of his life!

Ken Moses was replaced on the Board as Technical Director by Bert Wheeler, and his post as Group Manager for Notts Group was taken by John Longdon.

*Speech to the North of England Institute of Mining Engineers at their annual dinner/dance, November 1992*

This event took place a few days after Ken Moses' funeral and it was an occasion tinged with sadness.

'Mr President, Your Worship, distinguished guests, Ladies and Gentlemen, it is an especial pleasure for my wife Enid and myself to be here with you tonight. I suppose you are all thinking, 'I'll bet he always say that wherever he goes.' But let me assure you that the northeast — and I include north, and south, of the Tyne in that category — is held full of memories for Enid and myself and our family. Right through the sixties and seventies we spent two holidays each year usually camping at Embleton. The children acquired immense resistance to cold, particularly to their feet, which they got by extensive paddling in the North Sea.

I have to mention that, as a Yorkshire man, there have been times when I went down Durham pits, and sometimes even at Tursdale workshops, when I have needed an interpreter so that I could understand what the men were saying. But the manpower in the pits and the workshops were exemplary in their workmanship.

But, Mr President, there is no Branch of the Institution where one captures the sense of history of the mining industry like this one. In the library and lecture theatre in the Neville Hall building, I certainly felt that I was walking in the presence of the great and good of this industry.

The impact of the coal industry in the northeast is historically very impressive. We have done an exercise to determine the total output of the northeast coalfields since they started, up to the present time. That figure is a staggering, 5.262 billion tons. If all that coal was put into the standard 30-ton railway wagons, the length of the train would be 946,242 miles long, which is 2.2 times the distance to the moon and back.

Mr President, what an amazing achievement. But the northeast was also the development home for steam engines and the railways. And it was a major force in the expansion of the iron and steel industries.

The northeast coalfield has another advantage. There are plenty of reserves out under the North Sea, and working those reserves does not cause subsidence problems. The fish in the North Sea will not form a pressure group if the sea gets three feet deeper.

There is no problem, Mr President, in markets for the coal produced in the northeast coalfield. Coal has always travelled from the northeast by sea, and it is competitive to go into German power stations. But the European playing field is not level, and I see no chance of the German electricity industry using any UK coal.

We have to face the fact that, in business terms, the playing field is not level either.

Mistakes have been made in all the privatisations, — gas, telecoms and electricity generation — which have impacted on the market for coal.

While you will have heard of the 'dash for gas' in the past, you may not have heard of what I call the great travesty of financial justice which has applied to the mining industry. The mining industry has fought back long and hard, since the strike in 1985, to increase its efficiency and reduce its costs. In real terms, the industry reduced its costs to the extent of £1,411m last year. Where has that money gone? Have you seen it as a reduction in your electricity costs? The truth is that your electricity bills have increased by substantially more than inflation since 1985. It has been a convenient way of shuffling the financial figures around to ensure that the twelve electricity distribution companies, the two generators and the National Grid, increased their profits. Of the total of £2,900m profits made last year by the electricity industries, 47% can be ascribed to real cost savings in the coal industry.

Mr President, I have concluded that there is no logic in the present situation. It makes no sense for the coal industry to be sacrificed, but it is in the unfortunate position of being a pawn on the political scene. But sacrifice of the industry is a real risk at the present time.

What about the Institution and its North of England Branch? You must 'hang in there' as long as you can. I can see no other option for the time being. It might all be a tragic mistake, and the country might want the coal industry again. If it does want coal again, you have to be there to provide the professional focus for any revival.

So, Mr President, as never before, the North of England Branch is like a lifeguard and a coastguard, flashing the warning light but being ready to mount a rescue if needed. A most important role.

Mr President, it has been a delight for your guests to join you on this occasion. We thank you for your excellent hospitality and we hope that you enjoy the remainder of the evening.'

### Aspects of retirement

There were details to agree with British Coal as the date of my retirement grew closer. One aspect was that I selected a different way of receiving my pension payments compared with the normal practice. Most retirees opted to take their lump sum as a single payment, tax free, on retirement. My younger son, Adrian, suggested that the British Coal Pension Fund would likely be better at investments than I would. He drew up several assessments and charts that confirmed that, if I was to live seven years, I would then be much better off financially. I took his advice and took most of my lump sum as pension. This was the best financial decision I ever made in my life. Shortly after I retired, the Pension Scheme had a period showing a significant surplus. As a result, my bigger pension became even bigger. Retirement also paid off the outstanding loans for Vissitt Manor which had been provided by the Coal Board in 1977. We were thus free of debt. I had always stressed to my children that if they were free of debt, they would be rich. I also arranged to buy my car, a Ford Granada, from the Board. I did this through my former colleague and friend, Martin Shelton. He was Secretary of the Board at that time. The car had a magnificent sound system and I recall a time when there was a hold up on the A1M and I was listening to the slow movement of Rachmaninov's second piano concerto played by Vladimir Ashkenazi. It was so beautiful that I decided that I did not care how long the road would be closed. Surely this was a sign that I was getting into the right frame of mind for retirement.

The other major decision was what I would do in retirement. My colleagues looked at my varied experience in the Coal Board and suggested that I would be an ideal consultant for many companies associated with the coal industry. Enid and myself discussed this question and reached the decision that I would not take on any paid work, but I would do voluntary work in different aspects of the community. I was approached to undertake consultancy work and the people involved were surprised to have their suggestions declined.

### *The investiture*

The investiture was a memorable day, particularly as it was shared with me by Enid and my mother. Enid had been asked to ignore the splendour of Buckingham Palace when the doors were opened and instead head straight for the presentation hall and get seats near the front. When I eventually got to meet the Queen, I saw that Enid and my mother were in a prominent position within about 18 feet of the Queen.

*A very proud moment with my wife and mother*

All the recipients were trained in exactly what would happen and what they should do during each part of the procedure.

When I met the queen and she fixed my award, she noted that I had received it for my work in the coal industry. She commented on the important talks currently taking place about the future of the industry. I was able to say that I hoped they had a good outcome for the industry.

After the investiture, it was a tradition for the Coal Board to host the recipient and his wife to lunch. This was a very pleasant occasion, among colleagues and friends of many years. The Chairman made a presentation to me of a silver oil lamp inscribed, 'Presented to Trevor Massey by Members of the British Coal Corporation in appreciation of his distinguished services to the Coal industry.' It is on display still, alongside the oil lamp that my grandad used when he worked in the pits as a miner.

While Enid and I were at the lunch, my mother had sandwiches served to her in her room at the hotel. The hotel staff treated her very well as a distinguished old lady, and she spoke of their service to her with pride. In answer to a 'phone call later that evening about her day at the Palace, she said it had been wonderful. 'I'll tell you all about it when I see you.' She did indeed tell them about her memories of her day at the Palace — quite a few times in the future! She lived a further 12 years, and she was very sensible, near the end, to ensure that her assets were transferred on to her grandchildren.

### A quiet goodbye

I drove away from British Coal without any fanfares and with no regrets. It had been a fascinating and varied journey, from a humble start to a satisfying conclusion. My last eighteen months, in the Supply and Contracts Department, had seen a major restructuring exercise initiated and completed which reduced the personnel by nearly 40% with a similar reduction of costs. Also, the structure of the department had been simplified.

I felt confident that retirement would provide me with an interesting and varied life.

British Coal sent Enid and myself onto a retirement course at Leeds Castle in Kent. There were presentations of aspects for retirement on health, finances and planning that were helpful, but it was on another opportunity of retirement that had a bigger impact. We saw information about a holiday company, ATG of Oxford, that was news to us. They organised up-market walking holidays in Europe. On each holiday the ATG two-person support team, consisted of a manager and a guide, who led the walk each day. The luggage was moved by the manager each day to the next accommodation on the route. The manager also arranged a picnic lunch each day made up of local produce and local wine. The afternoon walking was undertaken powered by an alcoholic haze! The evening meal was always at a first-class local restaurant. The standard of the whole experience was excellent and the numbers in each tour were usually around fourteen. With our gardening commitments at Vissitt manor, (see below) our requirement was for more ATG type holidays in the winter months. They did expand into overseas trips and we did tours in Jordan, Namibia and three different ATG tours in India. We also met some interesting people over the years on these tours.

I was hopeful that retirement would provide me with an interesting and varied life and that Enid and myself would not be bored. My career in the coal industry had been a success; could retirement be just as good?

# Chapter 19

*Scenes from retirement*

### The Institution of Mining Engineers

In January of 1993 my first task was to hand over the Presidency to Albert Tuke.

Albert was a respected member of the Institution and was known for his management style of asking questions until he was assured that decisions made in the Institution were correct and effective. As immediate Past President I remained active in the Council and I was pleased to work with Albert during his year of office.

Quite soon we were involved in a significant issue with John Bourne, the Secretary of the Institution. He raised with Albert a problem that he had with his pension. He had a pension from his time in the Coal Industry and he had been accumulating an additional pension during his time as Secretary of the Institution. He had found out that the two pensions sums added together exceeded the total allowed by Inland Revenue and he would be liable for some tax. He requested that the Institution would pay this tax.

We all knew that John had expensive tastes. I was shocked in my presidency to find that if he had a meeting in London the total costs for the day added up to £500 which were charged to the Institution. He travelled by first class rail, he always took Ann his PA with him, and they had lunch at the Goring hotel. There was no way Albert was going to agree to the request until the issue had been fully investigated. The findings were very significant as, over the years since John became Secretary, he had transferred well over £300,000 into his and his secretary's pension funds. This fact had not been noted by the auditors as it was not shown as a separate cost heading in the accounts.

We took advice from a British Coal solicitor. He suggested that, as a charity, we should try to get the money returned rather than pursuing any criminal actions. Thus, began a long series of meetings with John and his secretary to seek their agreement for a significant part of the money to be returned to the Institution. John was reluctant to agree to this plan and Albert was frustrated by the lack of progress. Albert was succeeded by John Longdon as President. He took a very patient approach and I joined him in a series of meetings slowly tightening the screws. In the end over £200,000 was returned to the Institution. I was involved in finalising John's PA's pension. She was very happy with what she received as a starting pension and it had an escalation clause providing an annual increase of 5%. Another consequence of this incident was a change in the auditors of the Institution.

It was also when John Longdon became President that I took over from John Mills as honorary Treasurer of the Institution. This was to last for 14 years and during that time there were three major mergers of institutions in the mining and minerals fields. Each merger involved detailed discussions with the two bodies involved. There then followed the preparation for a new Charter and Bye Laws and the details of these had to have the approval of the clerk to the Privy Council.

The first merger was between the Institution of Mining Engineers and The Institute of Mining Mechanical and Mining Electrical Engineers.

The second merger was between the Institution of Mining Engineers and the Institution of Mining and Metallurgy.

The third merger was between the Institution of Mining and Metallurgy and the Institute of Materials.

The final title is the new combined body was the Institute of Material, Minerals and Mining (IOMMM)

As treasurer I was involved in these mergers which were important financially for continuation of the technical bodies concerned. The discussions demanded patience and flexibility to cover the wide range of the different members and their interests.

As Honorary Treasurer to the new institutions I saw my role to be the person who asked the questions (particularly financial ones) that the chief executives at that time did not wish to answer. After all, as a voluntary officer, one had no need to couch any comments with caution for fear of reducing one's chances of promotion or reduced salary. The society was getting experience and expertise free, a bargain price! All charities and voluntary organisations need someone to ask the difficult questions that lead to the avoidance of future problems and inefficiency.

*GP Fundholding.*

It was early on in my retirement in 1993 that my daughter-in-law had a quiet word with me about the North Thoresby GP practice where my son Brynnen was a Partner.

"They are going to ask you to negotiate the contracts with the hospitals for their Fundholding Account. They want to continue to concentrate on being doctors and not have their time diverted into negotiations with the hospitals."

The large rural practice had over 9,000 patients and they worked from two surgeries, one in North Thoresby and one in Holton-le-Clay. They had a dispensary to issue a significant proportion of their prescriptions. They had a Practice Manager who would manage the day to day operation of the Fund. She was a woman of experience and status in the operation of the surgeries, but she had no affinity for the cut and thrust of negotiating contracts.

I regarded this as an interesting diversion in my early days of retirement and agreed to take on the role with suitable payments for my travelling expenses to the meetings. Having examined the fundholding allocation, which was in the second phase of the NHS system, I decided how it should be managed. I agreed to meet the partners every three months to review the effectiveness of the system and the spend against the phased budget. I suggested that the contracts with the hospitals should be in two phases, April to December and then the final quarter January to March. If the spending on drugs was reduced in the first nine months of the year, additional procedures in the hospitals could be built into the contracts for the last three months. The money was mainly spent between the two local hospitals, Grimsby and Louth. However, the patients at this practice were prepared to travel further afield. This meant the fundholding team had to deal with numerous providers.

The other significant benefit arose when we met the hospitals to see it they could treat more patients in the last three months of their financial year. The additional work was a major bonus for them and they were keen to take it on. From our point of view, we needed to see a reduction in the cost of each procedure for us to be prepared to release the money. There were moments of tension when I sat tight while the hospital negotiator did his calculations. The Practice Manager would never have faced this negotiating stance. The payback was very significant for the practice. For many procedures we got a supermarket deal; two for the price of one or three for the price of two!

There was never any major shortfall in the hospital treatment of the referrals from the North Thoresby practice. At the end of four years, my report showed that waiting lists were very low. At that time 60% of GPs were managing Fundholding Accounts. I suggested to the North Thoresby GPs that it would be sensible for them to take on one or two local single GP practices into their scheme. This was not accepted as they were sceptical that an outsider would be prepared to work to their practice discipline on hospital referrals. (When I attended the Partners' Friday meetings each quarter, it was interesting to note how some difficult patient health problems benefitted by the pooling of all the GPs' experience to reach an agreed approach to their treatment.)

Of course, the Health Service is a major political subject, so a new Labour Government closed the Fundholding operations as being too market oriented, and they introduced a completely new system of managing the money for hospital treatments. If they had spread Fundholding to all GP practices, the Health Service might be very different to what it is today. After all, GPs know the needs of their patients and they are also aware of which consultants and hospitals provide the best value treatments.

### *Vissitt Manor— The Good Life*

Vissitt Manor was fully established as a productive garden and living centre when I retired. It was there to exploit and enjoy during retirement, while ever Enid and myself were both reasonably healthy. Strangers who visited Vissitt Manor said it was a paradise in its isolation. Our answer was that its paradise was earned by committed hard work. What we did is best described under separate headings.

### *Compost*

All the garden was covered by a sub-soil of clay. Any goodness fed into the soil was not lost as it would be with a sandy soil. Making compost was a continuous process. We had a shredder which would deal with any product up to 2 inches thick. There were large areas of grass. The grass and shredding products were mixed in a series of bays. When a bay was full it was turned over into a new bay. We also added the deep litter from the hen hut which acted as a stimulant. The process generated much heat and it was a favourite site for the cat to sit on the plastic cover and revel in the central heating. Through the year a large containing bay of compost was accumulated, at least 40 large barrows full.

The garden had a good complement of trees throughout and there was the problem of collecting the leaves. We had a ride-on mower which had a large collection hopper at its rear. We collected the leaves with this machine and they were stored in a separate bay which had a volume of about 12 cubic metres. They were left in that bay over the winter and water from one of the roofs was allowed to fall on the leaves. This was shredded in the spring and it formed a fine product of leaf compost ideal for potting on plants. It was even possible to use the machine at low revs to gather leaves on top of gravel in the front drive. We never composted potato plants as we did not want to risk getting blight in the compost.

In the later years we had applied so much compost that we had to raise the paths at the side of the 4 feet wide vegetable beds.

## *The greenhouse*

The greenhouse was the source of much produce over the year. The first crop was lettuce in a part of the outside bed. We also grew a climbing French bean Gadafal Oro. This bean was a failure grown outside, but grown in the greenhouse it was prolific and produced fresh beans each day from late March until the runner beans were ready. That area was then set with a late crop of lettuce.. The beans were not available commercially, so we had to keep the seed from one year to the next. We also set early carrots in two deep pots filled with fine sandy soil. It was possible to get two crops in these pots each year. The main crop in the greenhouse was tomatoes, both red ones and yellow ones. There were over twenty plants and it was normal to gather them by the bucket full and the plants were a trained so that they could produce over 15 trusses. The tomatoes were juiced and frozen in rectangular blocks in one of the deep freezers. There was always the option of having tomato soup on the menu at Vissitt Manor! The final crop in the greenhouse would be two climbing cucumber plants in one of the beds.

## *Soft fruit*

There was a composite range of soft fruits many grown along fences. Raspberries, strawberries, blackberries, sunberries (a cross between a raspberry and a blackberry), tayberries, loganberries, gooseberries, blackcurrants and redcurrants.

The redcurrants were grown along a fence and Enid pruned any side shoot so that the plants were restricted to a height of eight foot and a length of ten-and-a-half yards. They were very productive and were resistant to any wind damage.

Treatment of the fruit harvest was to juice the different fruits and freeze them in blocks in a deep freezer. They could be converted into jams and jellies later in the year.

We also had a rhubarb area. As we were within ten miles of Wakefield, the noted professional area for rhubarb production, we did have successful crops.

### Vegetables and fruit trees

The vegetable area of the garden changed from year to year. I made a plan of what was grown where each year and there was an aim to rotate crops. The strawberry bed moved over after three years and it was followed by potatoes. We were self sufficient in potatoes and we grew two varieties "Charlotte" and "Desire."

In the garden I said to the plants that I could not guarantee them sunshine, but I could guarantee them that they would never lack water. The large well under the back garden held many thousand gallons of water. I had a pump in it and filled containers in the greenhouse and in places within the garden. The well never ran dry and I was able to water the garden to get crops of big vegetables.

There was a good spread of fruit trees. Apples (both eating apples and cooking apples), pear, damson, and Victoria plum. We did set a few more fruit trees to get newer trees.

Each year the church in Barnsley had a plant sale. Enid bought hundreds of small flower plugs which we were able to pot on with all the compost we had. They became plants for the gardens of Barnsley. At the same event we sold jars of jam and honey.

## Hens

We had free range hens in an area adjacent to the bee hives. I bought the hens, when they were at point of lay, from a man who lived in the hills near Penistone. He worked for the CEGB and was a cunning salesman. After we had seen the hens that were for sale we always sat and had a chat. We discussed a range of subjects, world affairs, UK politics, his holidays abroad etc. He then asked how many hens I wanted and then he stated the price. They were expensive, but it was not possible to pull out of the deal after all the discussions! The hens normally settled down quickly, but on one occasion when I went out to check them at night fall, instead of being in the shed they were all nesting up a big apple tree. We had to climb up and recover them and put them to roost in their shed.

One day I got a call from my nephew Robin to say that the salesman had died on his way to Leeds market. How was I going to safe-guard the future supply of hens? We visited his house and bought two of his cocks that we could run with the hens. We then got a small incubator so that we could raise our own chickens. It was a minor miracle to see the chicks at birth forcing their way out of the egg shells. Enid was not keen on the cocks that we kept. They were proud and dominant over the hens. On a few occasions a cock butted Enid at the back of her legs when she was feeding them with her back to him! The hen food, which they had each night, was boiled up household waste mixed with hen meal. It always smelt very appetising. We also got corn from the farmer which we spread in the grass each morning.

## Bees

Enid had kept bees with her father since she was a teenager. When I retired I said I would play a much bigger role in the beekeeping. We had bees in three locations. 2/3 hives at my mother's house near Barnsley; 2/3 hives a Womersley, near Pontefract, which Enid's sister, Joyce looked after. At Vissitt Manor we had 6/7 hives. We never moved the hives to other sites, and we ceased to collect any swarms, other than our own. This was to avoid picking up the Varroa mite which was affecting bees in the UK at that time.

I made some spare wooden hives in the workshop so that we could expand our holding.

We were members of the Wakefield Bee-keepers Association. We attended their meetings and benefitted from the many years of experience those bee-keepers had. At that time there was interest by some members to import Queen bees from the continent on the basis they were very good. One Wakefield bee-keeper said: "There's nowt like Yorkshire bees for working in Yorkshire." We agreed with him and remained faithful to our Yorkshire stock. There were visits to the apiaries of the members during the summer. We usually had them at Vissitt Manor once each year. I think that Enid's sandwiches and cakes might have been the real attraction!

Tricks of the trade. Keeping the bees alive through the winter meant protecting the hives from mice and wrapping them up to keep them as warm as possible. In the winter the bees gathered together in a tight cluster for warmth, but they needed some food. This could be honey left in the hive in the autumn or the hive could be fed sugar syrup. One of the early crops worked by the bees was oilseed rape. There are many millions of flowers in a field of rape. So, we always fed syrup to the hives early in the year. Plenty of food would encourage the Queen to start laying eggs. This would give a good stock of flying bees when the fruit blossom was in flower just before the rape. The rape honey quickly sets in the hive like concrete and it needs to be extracted as soon as it is ripe. Even the bees cannot use rape honey as food if it is left in the hive. Obviously, the bees also work other fruit crops through the year. One late crop was willow herb, which was prolific along an old railway line within flying distance of the bees at Vissitt Manor.

On a hot summer's day, with a full flow of nectar, it is an amazing site to see the hives at peak performance. The bees come back to the hives with their pollen baskets fully loaded and, at times there is more traffic than at Heathrow airport. So they have to circulate before they can land at the hive.

Harvesting the honey from the hives was always organised as a full family process. Two people would have the job of taking the supers of honey off each hive and replacing them with empty supers. A full super of honey weighs about 30lbs so it is heavy work. The skill is to sweep off any bees on the frames of each super as they are a nuisance if they get into the house where the extraction process is taking place. We also kept a record of the number of frames of honey produced by each hive.

*Collecting and hiving a swarm*

*Two young helpers doing a quality test on the honey!*

The first job of the team on the extraction process is to use a hot knife to cut off the wax seal over the honey cells. These cappings are kept and later squeezed through a bag to extract the honey mixed in them. The frames are then spun, four at a time, in an extractor to get the honey out. Initially the rate of spinning has to be controlled so as not to damage the frames.

When the extractor is full up to the level of the spinners, the honey is then poured through a sieve into another container; the filter takes out any bits of wax or bits of bees and the residue is pure honey. It was then transferred into special buckets that each hold 30 lbs of honey. Fortunately, there was plenty of spaces in Vissitt Manor to store the buckets until the honey was required by a range of customers. The production figures were normally over 1000 lbs per year, but in the peak year the output figure exceeded 1600 lbs.

Two special memories of bee keeping. It is hot work in full protective clothing attending to the hives. Enid always ended a session by taking off her gloves and turning them upside down and pouring the sweat onto the floor. The other memory is of hiving a swarm. A ramp is placed in front of the empty hive, sometimes covered with a white cloth, to make it easier to see the queen. The swarm is poured from its box onto the ramp in a random heap. Then a few worker bees at the front will enter the hive to do an inspection. If they are happy they come out and give a signal which is passed down the swarm. All the bees then form an orderly queue and march into the hive. It is like a scene from the film "Quo Vadis." Another amazing example of worker control by bees.

Within an hour the bees in the new hive will be flying out to collect nectar for their store in their food bank.

*A new church*

Two Methodist Churches were within a third of a mile of each other on the west side of Barnsley; Old Town Methodist Church and Huddersfield Road Methodist Church. My father died in 1973 and there had been discussions about a merger of the two churches in his time. The discussions always failed because no one would agree which Church should close. In 1997 four members from each church met and I was one of the members from Old Town church. We reached an agreement and its main clauses were as follows:-.

1. Neither church would be closed.
2. An assessment would be made on who was using each building for church and also for community activities. The prospects for more users in the future would be assessed.
3. Initially worship would concentrate at Old Town Church but there would be some services at the Huddersfield Road church.
4. A plan would be prepared to agree modifications for one of the churches so that it was suitable for church and community use into the future. If no such plan was feasible a new church should be considered.

5. There should be an agreed forward plan by the start of the new millennium, January 2000.
6. Both Church Councils would meet on the same date and agree this plan.

The agreement was accepted by both Church Councils. The merger took place with the positive support of the membership.

Both buildings were in use by community users with a spread of activities. Clearly there was the need for enhanced and more up to date facilities in the final design. The options were examined. The Old Town church was surrounded by buildings so there was no room for expansion. The Huddersfield Road church did have land around it, but the existing building was not feasible for modification and extension. The report went back to the Church Council that the only real feasible option was to build a new church on the Huddersfield Road site.

This was accepted but there were some older members who were reluctant to forfeit what they had at Old Town, for an unknown new building. Their question was, would a new building result in a positive religious society?

We were fortunate to have a member of the church who was prepared to finance a proposed design for the site by an experienced architect who had designed new church buildings.

To achieve the size of facilities desired it was clear that the building would need to be at least two storeys. This was achieved by removing a large volume of ground that the existing Huddersfield Road church was built on.

Certain parameters were agreed with the architects for the design for the church.

1. It would have a Sanctuary that was always retained as a Sanctuary.
2. It would not have a pulpit, but it would have a stage that would be adapted with lecterns etc. for worship. (The initial size of the stage proposed was too small. I was insistent that the stage should be big enough to take Grimethorpe Band in concert!)
3. The seating in the Sanctuary would be fixed in the form of benches.
4. There would be an Upper Hall which would be separated from the sanctuary by moveable partitions which could be housed in special units

at each side of the hall.
5. Adjacent to the Upper Hall would be a kitchen for serving food and refreshments.
6. Behind the Upper Hall was a corridor leading to two offices and a bay of toilets.
7. The main entrance was at the upper level in the middle of the building giving access to the Upper Hall and Sanctuary. The church and circuit office was provided at the side of this entrance.
8. There was a meeting room in the attic, ideal for quiet meetings.
9. There would be a lift to operate between the street level and the upper floor which would provide access without mounting steps.

At the ground floor level (street level) there was a large hall, ideal for children and youth work; one small office; one small kitchen and toilets; another room which became an ICT centre run by Barnsley College. (The church got a grant to support this, but they had responsibility for maintaining the bank of 13 computer work stations.)

Quite simply, in total, the church had a sanctuary, two halls, 4 meeting rooms and an ICT centre.

The Sanctuary and Upper hall could be combined to seat an audience of 430 people for concerts. It very soon became clear that the acoustics of the building were excellent.

Having got the outline design there began the negotiations with the Church Council to get agreement to proceed. There was fear that the financial cost of the scheme would be too high, as the initial cost estimate was over £100,000.

It was at this stage that the minister, Rev Derek Hinchliffe suggested that we should invite two other churches to join the project, Pogmoor Church, which was very near Old Town church, and the Barnsley town centre church, Pitt Street. My reaction was "Best of luck mate" as I thought the scheme was against the scepticism prevalent in many Methodist churches at that time. Pogmoor declined, stating that they had a small number of older members who wished to stay together until they died, when the church could be closed. At Pitt Street, Derek's letter was read out to the

Church Council members at the end of a meeting that had discussed several problems but not identified any solutions. They eventually replied stating that they were willing to join in the project provided that they could be equal partners. That was the start of letters and discussions to clear up the position. We were very happy for them to join us as equal partners going forward. We could not agree to going back to the starting line as the design of the new church was agreed. We knew that Pitt Street had a single building that was adapted daily for use by church and community groups. There was also a view that the Barnsley circuit had faced declining membership for years, so there was no need for a new church. There were many hours of discussions with their members by Leslie Newton, who had become the minister for the church, and myself. It took several months before the Pitt Street members came on board. However, they proved to be a very positive influence, particularly when the church went into use.

There was continuing action to gather together additional funds from other sources. Key elements would follow from the sale of the Old Town site and the Pitt Street site. Negotiating with the financial headquarters of the Methodist Church in Manchester was not easy and fell as a task to Leslie Newton. The scheme needed a significant proportion of the money raised from the sale of these churches to make the project viable. However, Derek Hinchliffe had negotiated a deal with a member of the church to underwrite a percentage of the total costs of the scheme; but we could not disclose this to the congregation. That person did remain in the background of the project, but he did say that he had sat in the seating in some American churches where the seats were so comfortable that he did not want to stand up. That settled the decision on the seating for the sanctuary and the upper hall, and it proved a very good one. Leslie Newton placed an order for the seating with Sauder Company in Ohio USA. He did all the communication for this contract by e-mail. The contract terms required us to send a container to their site by the date of manufacture. It was also our responsibility to arrange the container's shipping across the Atlantic.

The Church set up a Project Team to manage the scheme chaired by Leslie Newton. It also set up a company, Emmanuel Methodist Outreach

Ltd, which handled all the finances of the scheme. The company was also chaired by Leslie Newton.

The contract to build the church was let to Saul Construction, a building firm from Hemsworth. The site was managed by John Saul, the son of the owner of the company. The choice of Saul Construct turned out to be a very good decision. Their craftsmen worked to very high technical standards, as shown when they rejected some deliveries of bricks which were slightly warped. John Saul spent a lot of time on site and he insisted that the men continually worked at a high rate of production. I acted as project manager for the church and attended the monthly meetings reviewing the progress of the contract. The architect's young man on site had a very high opinion of himself, but between John Saul and myself, we brought him into line and got him to resolve any architectural issues very quickly.

The foundation stone for the new church was laid by my mother assisted by a young boy from the Sunday school. The final design also included links to the past. The foundation stones of the three churches; Pitt Street; Old Town and Huddersfield Road were retrieved and built into a wall of the building. Also, the steps from the public house in Barnsley on the top of which John Wesley had preached on 30th June 1786 were re-built as a feature at street level outside the lower level of the church.

The new building grew in size but no one from the church membership had access to the site in the construction phase. Attention was focussed on the opening ceremony on 23rd March 2002 when a special service would take place attended by the President of the Methodist Conference, Dr. Christina La Moignan, who would open the building. It was going to be a very tight schedule to complete the building for that ceremony.

On the Saturday morning, a week before the opening, the members of the Church Council were taken around the site. They saw the size of the building and its facilities. But the state of the installation that morning really shocked them. There were electric cables hanging down in many areas; most of the walls were still to be painted, and there were no floor coverings anywhere. On the outside there was open earth where there should be a tarmacked road and car park. The Church Council members realised it was a big building, but they were certain there was no way it could be ready in one week for the opening ceremony.

But in that week there was a miracle, as hosts of men in different disciplines worked together to achieve an impossible task. And they did achieve the impossible. During the week, I was there with John Saul trying to manage the different jobs. On several occasions I suggested to John that some work could be delayed until after the opening ceremony. His reply was always, leave it to me, I'll fit it in. And he did fit nearly everything in. The tarmacking of the access road and marking of the car park was done on the Friday. I have a picture in my mind of John's father, the Chairman of Saul Construction's, hand-sweeping the car park on the Saturday morning!

Internally, the floor coverings and carpeting were done early in the week. Painting and wiring up the electrics was on-going all week. I had arranged for my Godson, Robin, to have a team of young farmers to unload the seating for the sanctuary and to build up the bench seats. I was asked what time I wanted the container lorry to arrive at the site on the Tuesday morning and I gave a half-hour range of time. "No, I want an exact time, please." I stated 8.45am. Exactly on time a large articulated lorry parked outside the church. We were given two hours to unload the container. The muscular farm lads sorted it out in under an hour. They then got themselves organised to build up the bench seats. There were some slight sizing issues with the shelves for the hymn books, but Robin sorted that out.

The electronic organ was installed on the stage and its range of different sized speakers were housed in a storage space in a unit at the upper front left of the sanctuary. The very large bass speaker just sneaked into the space.

The only thing that was not completed was the boiler plant which provided under-floor heating in the sanctuary and upper hall and the radiators in the other rooms. Instructions were given to the preachers that we needed 'hell fire' sermons to keep the congregation warm. (The insulation of the building is very good. Experience showed that it got too warm during big concerts with a full audience and performers; on one or two occasions persons on the stage had fainted from the heat. So, it is normal practice to turn the boilers off for those occasions, even in the winter months.)

*A painting of Emmanuel Church presented to me when I left to move to Harrogate*

*The view from the stage of Emmanuel Church prepared for a concert performance © Roy Lloyd of Barnsley Photographic Society*

The opening ceremony took place successfully and memorably, and few people were aware of the immense amount of work done in that final week.

John Saul and his firm did the church proud.

From the outset people's reaction to the building was extremely positive. I saw one of the older members of the Old Town church, who had always been against the project, heading towards Enid after the opening ceremony. Oh dear, what was he going to say? He said, "I have been dead against this scheme from the first time it was raised. I have to tell you that I was wrong. It is magnificent." What a relief, and a pleasant surprise.

The financial outturn for the project was also satisfactory. Many visitors to the church, while being impressed by the facilities, asked how much debt it was still carrying. The answer was zero. In fact, there was over £100,000 unspent at the end of the project. The company Emmanuel Methodist Outreach Ltd was closed and the surplus was handed back to the church, but ring fenced to cover any major building costs in the future.

The other comment was that the building was too big. The challenge to all of us that were involved in the project was to roll our sleeves up and make it work. As minister Leslie Newton set out to grow the congregation to fit the church. When he left in 2006 he had achieved that objective.

## *Emmanuel Church in operation*

The religious side of the church settled down very quickly. New members joined, attracted to the modern facilities. Some local preachers were apprehensive as there was no pulpit. The width of the sanctuary and the fact that they were wired up with a roaming microphone, encouraged preachers to leave the lectern and wander from side to side on the stage to focus their preaching to different parts of the congregation. The control desk gave a full range of options for sound and lighting and the use of a projector and screen. The banks of lights allowed performers on the stage to be highlighted during concerts.

Eventually a standard pattern of three services each Sunday was adopted. A 9-30am service in a traditional format started the day. At 10-45am it was the main family service which divided part way through to allow young people to go out to classes. At 7-0pm in the evening was a service with a focus on an evangelical format with music and a singing group.

Over a few months it became clear that many groups and societies wished to use the building. We required a caretaker/manager to be on site who could help to co-ordinate the use of the building. For a short period, we had one person in that role, but he moved on. We then obtained the services of a craftsman made redundant from British Steel. He was excellent and was able to service the equipment in use in the building. He has had a major role, along with a team of church members, in developing the use of the building for multiple community use.

Additional facilities were added to the building. A church member designed, manufactured and fitted stained glass windows for the front of the sanctuary. Another musical member of the church negotiated the purchase of a Yamaha grand piano to take its place on the stage.

There was a need to have an efficient booking system for hiring the building. We settled on a computerised system which was available on the internet for anyone to see what was happening each day. It was set up by my son, Adrian, using an adaption of the MIT in America programme for use of their university facilities. He later wrote an additional programme which identified the income from all community users and it displayed the total earned each week. Access to this financial information was limited to a very few officials of the church.

In the early days a fundamental question had to be settled. Some church members suggested that church users should have priority use of the facilities. I had to explain to them that Emmanuel was designed to be a church and community building. It was the same for everyone; and could only work on a first come first served basis. As we progressed some societies were making bookings for two years forward to retain their slots. Church members, like outsiders, became adept at checking the forward bookings which were on the internet, to see where there were spare slots. It was not uncommon to have five of the seven halls and rooms in use on one evening. The financial return from the community use of the building was at such a level that the church effectively had free use of the building.

The use of the church as a concert hall quickly became established and the excellent acoustics were a delight for the performers. In time a full concert service was offered by the church. It included the provision of tickets for the numbered seats (430); concert programmes along with refreshments and ice creams in the interval. At these concerts the church would be receiving in excess of £500 for the night. There was a team of helpers who would supervise the whole event and, at the end, stack away all the chairs in the Upper Hall and return the building back to its normal format.

There was limited car parking on the site but there was ample parking along the roadside and also along a one-way street adjacent to the church.

It would be possible to write much more about Emmanuel and its expanding activities. But I think it is best concluded by what I said to Leslie Newton when he left in 2006.

"In 1999 Leslie you took over a church community hesitantly preparing to start on a journey towards a new church. You have led us on that journey and you have brought us through to a new church and to the promised land. And the promised land is more wonderful, more exciting and more challenging than we ever dreamed it could be. That is not just for the church members. Emmanuel is fulfilling that part of its mission statement that says, "reaching out to the whole spectrum of the community in service and evangelism, and ensuring the use of the resources for youth and community activities" In many weeks over 1000 people pass through Emmanuel. (In the month of December 2017, 3000 people attended concerts at Emmanuel) I ask myself why is it that the promised land is so wonderful Leslie? I believe the reason is that your ministry has ensured that the spirit of the living God is moving in this place. And that spirit is not just moving in this place on a Sunday; it there on every day of every week. So, anyone who uses the building feels it during their time there. The visitors who come here also feel it and they comment on the atmosphere of Emmanuel church."

## Writing

I had always like to write poetry and short stories even in my early busy life in the Coal industry. But when I became a colliery manager I accepted that my time for fiction writing was ended. A box was filled with the unfinished texts and my wife, Enid, mischievously wrote "Literary Masterpieces" on the box and it was stored away. There it rested for 22 years. Retirement seemed to be an appropriate opportunity to consider going back to some fiction writing. So, the box was retrieved.

In it was the start of a novel about a mining community in Yorkshire. There were 40,000 words of the novel and all the main characters were developed. I liked the characters who appeared realistic. It appeared to me as a challenge. Could it be worth finishing the book? I decided to take it on.

With my later experience in the coal industry, the plot became very different to what it would have been, had the book been completed in the 1960s. It became a long story and it was written slowly. By taking my laptop on our holidays it was written in many different countries of the world. The final scene in "Digging Deep," when the Dobson family were reluctantly leaving Upthorpe and the closed colliery, to go to new life in the Midlands, I tried to capture the pain of this time in their lives.

> "Helen and Roy Dobson with their two boys, Robert and John set off as a family on the car journey to their new home in the Midlands.
> "Why's Mum crying?" asked John.
> Robert, his elder brother, dug him in the ribs and told him in a whisper to "Shut up".
> Helen wiped her eyes, and when she calmed down, she replied. "I am sad, John, because of all the friends and people we are leaving behind. Just think, we won't see Tony, Michael, Dawn and Amy-Louise growing up. And I had so much more I wanted to do in the church. The last few years here have been as near paradise as we will ever experience. But it's all collapsed in the last three months.

*So now we will have to start and build it all up again."*
*"We'll help you, Mum," said Robert. Helen burst into tears again."*

That scene was written in the lounge at Hong Kong airport. The challenge in writing is to get the right words for any scene that stirs the emotions of the reader to laugh or cry.

It took a long time to complete the book. I passed it to an old school mate of mine Derek Hinchliffe whose father had been a miner. Derek gave me many helpful suggestions. Derek reviewed the book when he was ill and I was sad that he died just before I was able to give him a published copy. After Derek's work, the book was then passed to Judith Mashiter who had edited and produced two books which Derek had written. She decided that the scenes of the book should be dated and the whole story takes place in four years from 1964 to 1968. She also supported tabling a list of the characters at the beginning of the book.

Judith also came up with a range of titles and we finally settled on 'Digging Deep.'

It was published in 2012.

It has been very satisfying to have positive comments about the book particularly from people who had lived in mining communities in the 1960s. One woman said she felt she could become a colliery manager having read the book! Others were delighted to see the members of the Enclosure in the story. They remembered seeing similar groups of old miners in different villages commenting on the local pits. Some people said they shed tears for some scenes and one woman challenged me for killing off George Turnbull, the undermanager, who she saw as a hero of the book.

As there are now no pits left in the country, I hope Digging Deep will be around on the odd bookshelf as a social history book about a Yorkshire Mining Community.

My next writing project was to expand a female character who had appeared in a short story, Lucy Platt. She was a Yorkshire woman with a strength of character and purpose that I admired. The first chapter of the book started with the short story, but the main character of that story was her husband. A cousin of mine, who read the first draft, suggested that that chapter did not fit with the rest of the book. I agreed, so a new first chapter was drafted. The major theme is the development of a local bus company. I used the location and site layout which existed in a local bus company near where I was born to describe the development of Maxwell Motors, the bus company. Again, the book was passed to Judith Mashiter for her editing and help with organising the presentation of the book. Judith's genius was the photograph which she sourced to go on the front cover of the book. When she sent it to me I saw in the eyes of the photograph that same intensity of drive and purpose which I had always visualised in Lucy.

The book was seen by some as a story describing the development and growth of a Yorkshire family, and I heartily agree with that assessment.

I have heard descriptions by some writers of their emotions when they are putting down the words. Tears streaming down their faces as they type, but from their own feelings they know that they are using the right words which will impact on the readers. I agreed to do a last detailed read through of Lucy Platt. I was not very happy to take this on, as I knew there were a few places that I would have difficulty seeing the words as I would be mopping my tears with a big handkerchief.

I continued to write, and another novel was half completed, but I was encouraged to focus on my life, with all its twists and turns in the coal industry. The first draft finished in 1992 when I retired from my paid job in the industry. But my retirement has been longer than expected, now 26 years in 2018, and it has continued to be varied and interesting. So, a second draft has been prepared to complete this publication, with added scenes from my retirement years. After all, I have more experience of retirement than any other phase of my life!

There you are, A Life of Changes and Challenges. In retrospect my life was loaded throughout with good fortune and the support of a loyal family. It might even qualify to be termed a wonderful life!

# Appendix

*Cardiff, Thursday 19 March 1992*

'A Yorkshire man needs a letter from some dignitary which guarantees that he does not support any Rugby League Club and is thereby not a contributor to the export of Welsh rugby talent into the northern hills. If he does not have this, at least he should have a reference from the local church Minister which guarantees that he can sing either the tenor or bass line in Cwm Rhonda. Mr President, I have none of these references, but I do approach this meeting with an air of confidence for an entirely different reason.

I must dip into my past for this assurance. In West Yorkshire, in the Dearne Valley, the layout is very similar to the Welsh Valleys. Woolley Colliery, along with Woolley Colliery village, was just up the hillside from the river, and the railway in the valley bottom. Above the pit, further up the hill, there were two buildings: one was the colliery manager's house and the other was the village school. For three and a half years in the late 60s, I lived in the colliery manager's house and my kids went to the village school. One day my eldest lad, who was about 8 years old at the time, came home and said, 'Dad, I would like a word with you. I'd like to ask you a question. Why have I got two Welsh names? Why am I called Brynnen David? Why didn't you give me two Yorkshire names?' This was a very serious family situation, as you will appreciate.

I sat him down and said, 'It's like this, son. On the very obscure chance that I should become President of the Institution of Mining Engineers, and be invited to speak in Wales, I will need something to demonstrate to the

Welsh mining engineers, my respect and affinity to them. You are it! When they know that I've named my eldest son and heir with two Welsh names, they will give me a round of applause and welcome me into their midst.

Mr President, as I drove down from Yorkshire to Cardiff today there was little evidence of the coal mining industry. It would have been very different 80 years ago, just before the First World War. Then there were 233,000 men employed in the Welsh mines and the annual output was over 56million tons. Very impressive, but probably more impressive is to look at the total historical scene for Welsh mining.

Over 300,000 million tons has been extracted from the valleys of South Wales since mining began, which is well over 200 years ago. This is a lot of coal by any standards. To get some appreciation of this, if all that coal was placed in a standard 32-ton wagon of a merry-go-round train, it would need 93 million and 750 thousand wagons. Putting that into perspective, it would form a train 513,878.1 miles long, which would stretch around the globe nearly 21 times.

Mr President, these figures represent an awesome achievement by the miners of Wales. When one considers extracting 300,000 million tons from the Welsh hills and valleys I am somewhat puzzled that there are any hills left. Why isn't the Welsh topography now as flat as Lincolnshire?

At that time the total production in the UK amounted to 23% of the world production. Of that UK total output, 97 million tons was exported, which represented 55% of all the coal traded internationally. Almost half of the Welsh coal produced was exported and it literally went to every part of the world. The world and its industry were being fuelled by Welsh steam coal.'

The speech went on to reflect on the current projections for world coal consumption, which was expanding significantly. This was justified when the increases in population were noted. The world population was, at that time, expanding at the rate of 100 million people per year.

My speech then itemised the difference between the international scene for coal and the UK prospects. It noted the subsidies for nuclear power in the UK and the dash for gas.

'The most important point is to consider where the UK stands in the technical sphere of coal mining. Has the industry declined because we have fallen behind technically in the world coal mining scene? The answer to this is emphatically, no. In fact, in the UK we have had to solve technical mining problems which have not yet been met in other parts of the world. So, technically we are in a very advantageous position internationally. This is also readily demonstrable by the number of UK mining engineers now working overseas.

It is a major responsibility of the Institution and its Kindred Societies to maintain the morale and hopes of the young engineers in mining. They need to be encouraged to see their future on a much broader scale than has applied in the past. Theirs is a more exciting future than those of us who have been in one company, in one country, for over 40 years have experienced.

Mr President, these are very challenging and uncertain times for all of us. It is essential that the Institution and its Branches should be active and effective in keeping the mining family together and resolute to face the future. Your Branch is fulfilling that obligation and your guests, all of them, are appreciative of the opportunity to share this dinner with you.'

It concluded by me asking the audience to drink the toast to the Welsh Institute of Mining Engineers.

### Midland Institute of Mining Engineers, Sheffield, Friday 20 March 1992

'The Midland Institute of Mining Engineers has always been a problem when talking to people outside this locality. Where are the Midlands? For people south of Watford I have no doubt the Midlands must be somewhere around Milton Keynes. As far as they are concerned Yorkshire is very much in the far north.

I have the same problem with the terminology in the coalfield where the Groups have lost their links to Yorkshire. Mr President, I have to admit that I was much happier in the 1970s and 1980s when the four Areas were very firmly linked to Yorkshire. At that time, they were all very different, and it is worth pondering a little on those differences.

The North Yorkshire Area was clearly linked to Rugby League. I remember being in one pit and seeing a Rugby League forward holding up the crown of an arch girder while his mates set the legs. It was always denied by the Inspectorate, but I am sure that at Sharleston pit they had an exemption not to use horse-head girders if there was a Rugby League forward in the ripping team.

South Yorkshire Area, at that time, was the home of quality fuel. Any pit sending a train load of coal to one of the power stations was frowned upon as one of an inferior breed.

The Barnsley Area — and nothing could be more Yorkshire than Barnsley — was at that time the home of three centralised coal preparation schemes, designed to produce top quality coal. It is a sad sign of the change from those times when, at present, no coal is sold to British Steel and the house coal market is being eroded by imports.

And then there was the Doncaster Area. This was unique. The ultimate experience for the Mining Engineer. And many were thrown in to gain that experience. Any who passed through and got to the other side really did think they had reached the Promised Land. The uniqueness of Doncaster Area showed itself in many different forms. For instance, if someone gave an order, it was not a signal for action. It initiated the start of a debate. The labour force might not have had very productive miners, but they were the best talkers and debaters in the country. It led to frustration and some unusual approaches.

I am sure Jack Wood, when he was the Director of the Doncaster Area, who is here tonight, will not mind me passing on to you one unusual approach which he made when spirits were at a low ebb among the Doncaster Area management team. He asked Peter Cotgrove, who was the Chief accountant and a local Methodist preacher, to have a word with the General Manager in the Sky and see if He could help us to get more output. Whether it was a bad line or whether Peter got the specification too vague, or whether Upstairs they only deal in general terms for fuels, I don't know, but the result was an outburst of methane at Yorkshire Main colliery the week after, which was the biggest that had ever occurred in a UK mine. We had plenty of fuel, but it was the wrong sort.

The general effect of the Doncaster Area on Senior Management was covered by the medical term 'Doncastration.' This showed itself in psychological effects. One survey carried out among the wives of Senior Managers into the effects of 'Doncastration' found the following evidence. Senior Managers with 'Doncastration' become listless; they become restless; they can't sleep, and they are short tempered. Their performance in bed is much reduced.'

The speech then took a serious turn and discussed the unfairness of the current marketplace for coal, and listed various decisions and commercial practices that were detrimental to the coal industry. It identified, however, the potential for improved results in the coalfield.

'This coalfield is now the scene of one of the great commercial battles of all time. What is happening is that Alan Houghton and his team at Selby, Bob Siddall and his team in South Yorkshire, supported by contractors and suppliers, are battling to raise performance and lower the costs of production. As an industrial tussle in the marketplace, in my opinion, it rates with the great battles of the Second World War. You may choose your own example, but I compare it with the Battle of Britain or the defence of Stalingrad. I very much hope that the battle ends in victory.

As for the Midland Institute of Mining Engineers, the Institute provides a forum to meet and discuss the technology options for the industry. The meetings help to weld together all the parties in that body, which is bigger than the Institute and bigger than the industry — I refer to the mining fraternity. As President of the Institution of Mining Engineers I am deeply grateful for the support given by the Midland Institute to the Institution and I hope that that support may long continue. Mr President, gentlemen, will you be up-standing, and I give you the toast to the Midland Institute of Mining Engineers.'

## Mining Institute of Scotland, Saturday 21 March 1992

'As I am moving around the country as President, I am beginning to be aware of the strange looks I'm receiving from members of the Institution. The looks define an unspoken question, 'Is he here as a Mining Engineer, or is he here as a shopkeeper?' I have to say, Mr President, in coming here tonight I have my purchasing and stores hat firmly on my head.

One of the key roles in purchasing is to observe if there are any signs of a cartel forming among suppliers. I detected some signs of a cartel between the secretaries of the Mining Institute of Scotland, the South Wales Institute of Engineers and the Midland Institute of Mining Engineers when they invited me to speak in Wales on the Thursday night, Sheffield on the Friday night and here in Scotland on the Saturday night. Was this a test for the President? Would he use the same speech on each night? If he used a different speech it would be a test of his managerial ability to see that he ends up with the right speech on each night.

These three nights in the pop music world would be called 'a grand tour of gigs', while in the world of TV it might be compared with the Antiques Roadshow. Looking round the audience tonight, I am not sure which description would be the most appropriate.

I intend tonight, Mr President, to reflect on my latest role in the corporation as the shop keeper and buyer.

It is interesting to see the different approaches by different people when purchasing a product. The ladies present know the problem they have when the whole family go shopping for clothes for the kids. The mother wants clothes that are smart and serviceable, and reasonably priced; the kids want something which is casual, garish and colourful, which fits the latest fashion, and they don't give a damn about the price; Dad wants anything so long as he can get home to watch the rugby on the television.

There are similar diverse interests in British Coal. The key man in the chain is the colliery manager. He wants something, he is not quite sure what he wants, but somebody mentioned it to him in the Club — but he certainly wants it yesterday. The same product for the Mechanical Engineer has got to be specified in terms of reliability.

He therefore wants it made in inch plate, such that it will be unflinching whatever abuse it suffers in use. The Electrical Engineer just knows that no-one else will possibly understand what he wants, so he prefers not to give a spec at all, so he just asks for a black box. The Surface Foreman's view is quite simple. No matter how big the machine or piece of equipment, he wants it in a pack which can be loaded by a fork lift truck into a mine car. We now have a new person appearing — the environmentalist. He wants it in a colour to blend in with the environment, even if it is buried with dust and muck a mile underground.

Now the supplier adopts a most helpful role. He doesn't really want an order, he prefers a long-term contract for 20 units over 2 years, so that the colliery can just call one off on demand. He also wants payment up front, and he requires a clause in the agreement that he can alter the specification and up the price every Friday at 12 noon.

Now how does Supply and Contracts Department meet all these varying requests? It is brilliant the way the department, over many years, has set up a system whose procedures only they understand. The first thing to note is that the whole of the purchasing mechanism has been computerised. The codes to get into the computer are only understood by one or two young ladies in an obscure corner of Coal House. Supply and Contracts don't identify anything by name. Everything is given a vocabulary number, which has 8 digits, and an order number which has 6 or 7 digits. With so many numbers, the chance of getting a vocabulary number and order number right first time are slim.

To place an order, a whole suite of forms has to be signed by the appropriate person. Again, I don't think I am exaggerating when I say that very few people in the corporation understand who is supposed to sign which form. I preside over this service department, Mr President, and marvel at it as one of the great wonders of the world, because equipment and consumable items do in fact appear for the user at the end of the process.

Of course, what I have just described does not apply in Scotland. Mr President, your colliery stands alone and is serviced by Headquarters direct. My department staff are full of praise for the clarity and understanding there is between you and your staff and the products and services you need from Supply and Contracts. Scotland is different compared with the rest of the UK.

The Mining Institute of Scotland also differs from the other Branches of the Institution in the way it integrates opencast mining along with deep mining into the Branch affairs. Of course, opencast is the big player in this combination. It is interesting to note that, in the first ten months of this financial year, the output of Scottish opencast sites amounted to 3.5 million tons and that 24% of that coal was exported into England. I promise not to broadcast that figure, Mr President.

Opencast operations in Scotland are also noteworthy and different from their English counterparts in their ability to get planning permission for opencast sites in a structured way. Two new contracts have been let this year, and there are letters of intent for a further three sites.

The Institution of Mining Engineers wants to have active members in disciplines involved in the many facets of the business of mining and other extractive industries which applies in your Institute.

So, Mr President, I would submit that the Mining Institute of Scotland is in the vanguard of the changes necessary for the future of the Institution of Mining Engineers. Quite clearly, the Institute in Scotland is held in affection by engineers in many disciplines from other parts of the country. That is evidenced by the support for you and your Branch tonight.

Ladies and gentlemen, it is with pride that I ask you to charge your glasses and to drink the toast to the Mining Institute of Scotland, coupled with the name of its President, Mr Ramsey Dow.'

### Notts. & North Derbyshire Branch, 3 April 1992

'There is the breakdown of the communist world and the emergence of those countries and Russia to a regime of freedom. In parallel with that is the integration of the European states with the breakdown of barriers and national boundaries and the freedom of movement along with the freedom of opportunity for business throughout Europe.

'In travelling here tonight I crossed the boundary between York-man and Notts-man. It is the natural divide between the descendants of Viking man and Saxon man.

Mr President, I live in the heart of Viking country, not far from Grimethorpe and only a few miles from Barnsley, which could very well claim to be the capital of Viking-land. Barnsley suffers from an inferiority complex. It is not uncommon to see a hitchhiker on the M1 coming north holding up a sign which says 'anywhere, except Barnsley'. In a similar way you might see a train with a small notice which says, 'do not flush toilets in the stations, except in Barnsley.' Barnsley is not noted for producing famous men, but its most recent one, I suppose, could claim to be Arthur Scargill. Arthur has done for Barnsley what the Boston Strangler did for Boston.

There are attempts being made to lift the standards in Barnsley. A man recently signed in at a hotel in Barnsley and the receptionist asked him. 'have you got a good memory for faces?' He replied, 'Yes,' and she said, 'Oh that's good, because there is no mirror in your bathroom.'

Mr President, those little stories may give you some feel for Viking-land. Many of you might know that British Coal has decided to sell Coal House. For those of us who work there, it came as a bit of a shock and we had to think rather quickly. You will now find members of Supply and Contracts Department busking on Doncaster station to earn a few coins so that we can buy a marquee to put up on one of the roundabouts on the outskirts of Doncaster.

My instructions from the corporation is to lead the troops south; I interpret this as a suggestion that we should march towards the centre of the earth which, at the present time, seems to be Eastwood Hall. The fact that there is no room for anyone else there is just incidental.

Mr President, Utopia seems a long, long way off in the great uncertainties facing our business today. But Utopia was a long, long way away in the 1984/85 strike, when there was a very clear difference between the confrontational nature of the sons of Vikings and the flexible attitudes of the sons of Saxons. Many, many people in this country had then to be thankful that the sons of Saxons stood by their principles and insisted on working the mines to defend those principles. We have all got memories of that time and the fear and bitterness which ripped its way through our communities, affecting all our lives. One memory that I will always retain is meeting a young man underground at Thoresby pit who had continued to work. He was a special case because he lived near the pit and he had to walk from his house, through the picket line, to get into the pit each day. He also had to walk through the same picket line to get home to his wife and family at the end of his shift. The Manager introduced him to me with pride and I shook his hand, humbled at his immense courage and his bravery which he had to produce every single day.

Mr President, I fear for the future of this great industry. It seems that a few administrative decisions over the last couple of years have barred us from our rightful place as the main fuel source for electricity generation in this country. One also gets the feeling at times that, in the creation of the new businesses associated with electricity generation, there is a kudos and a mission to use any source of fuel other than British coal.

In these circumstances, Mr President, this great branch of the Institution has to continue to play its part as a forum for discussion and debate and for the dissemination of technical expertise among the mining fraternity. It has a major role to play in this coalfield in the future. Ladies and gentlemen, it is a very great honour for me to ask you all to rise and charge you glasses as I propose the toast to the Notts and North Derbyshire Institute of Mining Engineers.'

*South Staffs & South Midland Branch, 24 April 1992*

The final speech before the Annual Conference was on 24 April at the dinner/dance of the South Staffs & South Midland Branch at Penns Hall. As there were ladies present, it was an opportunity to have a little fun with the ladies.

'It is a delight to be with you tonight, accompanied by my wife. I can see that some of the ladies are showing sympathy for my wife who has to cope with a husband having to tour the country giving speeches as well as doing his job. My wife is pleased that I have joined the AA to gain support for my busy life. I can see that some of you are looking a little puzzled as to what the Automobile Association has to offer me. I can see that others are thinking, 'Ah, what he means by AA is Alcoholics Anonymous, he has a bit of a drink problem.' But again, you would be wrong, because anyone who knows me will tell you that my intake of alcohol is somewhat limited, so I don't need to be a member of Alcoholics Anonymous. What I mean by AA is Athletics Anonymous. This is a relatively new service which now has a branch in Barnsley. If you feel a jog coming on, you ring their number up, and they send a sexy young woman out to stay with you until the urge has worn off. So, I can assure you that athletics is not my forte.

I must say that having a woman sitting on the settee, drinking orange juice, might be a little off-putting for the wife of the household. My wife has taken a very positive approach to this situation, but she has modified her reaction to meeting new people. Whenever we are out in new company she always introduces me as her first husband. She has found that it keeps me on my toes and secures her options for the future.

Since I became President in January, it has been my lot to be giving speeches during an election campaign. Can you imagine anything more soul destroying? The audiences were saturated with politicians holding forth on a range of topics. I suppose it was all covered by the general term 'faith, hope and party.' Never before have I been so aware of the accuracy of some of the sayings relating to politicians. There is one saying that refers to the difference between an academic, a manager and a politician. The academic argues to a conclusion, the manager argues for a decision, the politician just argues. Another quotation which I think is valid says, 'It is easy to be brilliant if you are not bothered about being right.'

How does one distinguish between politicians? Another quote: 'the best lack conviction, the worst are full of passionate intensity.'

What do they really think? Are they like the Irishman who said, 'how do I know what I think until I hear what I say.'

Another factor is that political truth has a short time span. It is the supposed truth that wins the next vote or the next election. It can also be the truth which fudges a problem for the time being.

By contrast, Mr President, a true picture in the mining industry is not a snapshot at any point in time. It must be valid over months, or even years. Any Colliery Manager who leaves a pit, if he has left it right, must rejoice in the better achievements of his successor than he ever achieved.'

The speech then discussed the serious issues facing the industry and tried to bring them into context.

'It has been the misfortune of our industry to be subject to short term decisions in other fields over the past two years which are now having a major impact on our business and its future. It did not seem particularly noteworthy when the Minister decided to take the nuclear generating industry out of the electricity privatisation process. But in so doing he removed it from the competitive challenge and he gave it a base load role for electricity generation at a massive subsidy. This has, at one stroke, taken out more than 20% of our rightful market.

The decision in the electricity privatisation to encourage the regional electricity authorities to seek sources other than the main generators seemed sensible and innocuous. It led to the dash for gas. The gas power stations are now covered by long-term contracts, giving them base load generation for 20 years, and again taking over 20 million tons of coal generating capacity out of the market.

Mr President, I have certainly tried to demonstrate tonight how our coal industry has been handicapped by the political decisions in the recent past. The industry itself is in the twilight of its life but it has so much still to offer this country in the future. We also have those in our midst — young mining engineers and miners — who are in the dawn of their careers and who are seeking a long and secure employment in the industry. Mr President, there is a moral test on the new government in the UK on how it treats this industry and the people who work in it. I hope it passes the test. The impact will not just be on the mining industry, it will also be on the Institution and on this Branch of the Institution. Ladies and gentlemen, I would ask you to charge your glasses as I propose the toast to the South Staffs & South Midlands Institute of Mining Engineers.'

# Glossary of Technical Terms

**Organisation of the National Coal Board in 1951**

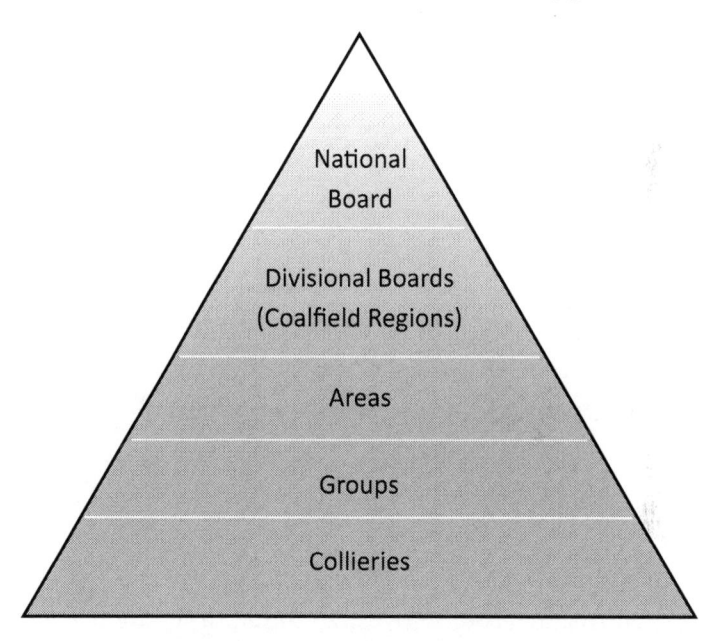

*Pyramid diagram (top to bottom): National Board / Divisional Boards (Coalfield Regions) / Areas / Groups / Collieries*

In 1951 there were eight Areas in the Yorkshire Coalfield Region. The N°6 Area had the mines divided into four Groups, each with a Group Manager and a small team of experts. Most mines had a Colliery Manager, but with the very small mines a Colliery Manager would cover two mines.

## Coalfield geology

As well as the coal seams the mining operations are affected by the adjacent strata. This was mainly shale beds but it could be sandstone, and occasionally there could be ironstone beds adjacent to the coal seams. The specific gravity of the coal is 1.3-1.4. The specific gravity of the shales and sandstones is 2.4-2.8. The specific gravity of Haematite is 5.2. The specific gravity of the rocks is significant as they have to be separated from the coal on the colliery surface. In some coal cleaning plants the mixed run of mine coal is treated in a unit which has a dense medium set at 1.4 specific gravity. The coal floats through on the surface and the other rocks sink to the bottom and are extracted.

## Cleat lines in a coal seam

Coal seams are not composed of solid coal, like concrete. Due to the pressure applied during the formation of the coal, in varying degrees, there are planes of weakness. It is these lines of weakness that are used by miners to peel off the coal from its solid state.

## Mine roadway

These are the permanent passageways retained after the coal is extracted to allow access to the mine production workings. In most seams they are formed by extracting shale above the coal seam to achieve a height which allows men and materials to be transported from the pit bottom to the coal faces. Permanent supports are erected, usually steel arches, to avoid the roadway collapsing.

## Cutter picks

These are the cutting tools used to cut the coal or dirt in the various mining processes. They are made of steel, for strength, to fit into the cutting machines. They have very hard tips to cut the coal or dirt; the picks wear down in time and need changing.

## Man-riding train

This is a purpose-built train with carriages to carry workmen from the pit bottom into the workings. It can be powered by a rope haulage or by a locomotive. In some mines, systems are available for men to ride on conveyor belts. The aim of all the systems is to get the men to and from their work, quickly and safely, without them using their energy by walking.

## Shift times

Mechanised mines are usually operated through 24 hours of the day. The dayshift would be 6.00am to 1.30pm, afternoon shift 1.00pm to 8.30pm and the nightshift 10.00pm to 5.30am. The gap between the end of the afternoon shift and the start of the night shift might be used for special maintenance work.

## Hand-filled coal face

These were the majority of coal faces when I started work in the industry in 1951. The coal faces could be up to 250 yards long, with a central roadway which carried the coal away from the coal face on a conveyor and provided a flow of fresh air to the coal face. At each end of the coal face would be another roadway which was used for access for the miners and to provide supplies. Those roadways also took the air which had passed along the coal face away along main roadways to the shaft and up to the mine fan to be exhausted to the atmosphere.

The method of working the coal was to undercut the coal seam to a depth of about 5 feet. An early type of coal cutter is displayed in schematic form, showing its three sections. The jib goes under the coal seam and provides the race along which the cutter chain travels. The machine cuts a slot under the seam about five feet deep. As the slot is cut it is usual to provide support with wooden wedges pushed under the start of the cut.

The coal face is then drilled to provide bore holes to take explosives that are fired to break down the coal. The miners then fill the coal onto conveyors and set wooden supports to hold the roof.

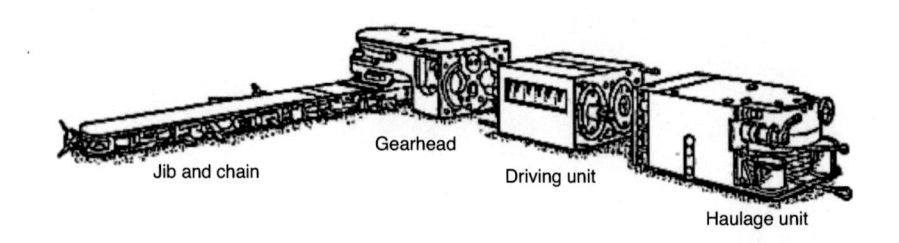

## A mechanised coal face

As illustrated, the machine cuts and loads the coal onto a panzer chain conveyor. The cutting machine (usually a shearer) rides on the panzer conveyor. The line pans of the conveyor have flexible joints so that the conveyor can flex and be rammed over when the coal is extracted. The armoured face conveyor was a German design (hence 'panzer' conveyor!) and it introduced the concept of prop-free front working.

Rams connected to the roof supports (chocks) are used to push the panzer conveyor over. The same rams pull the chocks over into the new position when the supports are lowered off from the roof.

All the workmen operate in the supported area behind the conveyor. The powered chocks are usually designed to give immediate support to the roof when the shearer has cut the coal.

## A ROLF coal face.

A Remotely Operated Longwall Face operates as described above as a mechanised coal face, but the movement of the panzer conveyor and the powered supports can be done remotely. A control panel in the mine roadway knows the exact position of the coal cutting machine and ramming over of the panzer conveyor and advancing the supports can be initiated by the man at the control panel in the roadway or can be done automatically according to the position of the shearer. These systems reduce the number of men on the coalface team. The technology is effective in seams which have reliable geological conditions.

## Retreat coal faces

For these coal faces, the roadways at each end are driven out to some predetermined position. The coal face is then formed by connecting the two roadways, and the coal is mined allowing the strata to collapse behind the coal face. In some mines the coal face is short, say 50 to 70 metres, and pillars of coal are left on each side of the retreat coal face. This is often to ensure better roof conditions or reduce surface subsidence. Retreat coal faces can be up to 250 to 300 metres long where the geology is good and there are low quantities of methane released as the coal is mined.

## A single-entry coal face

Single-entry coal faces were designed for a period at Wistow mine in the Selby Coalfield. A mine roadway was driven out to the boundary and then a 45-metre coal face was opened out. The roadway and the coal face were ventilated by an auxiliary fan from the main roadways. Special arrangements were made to ensure the coal face was clear of any methane gas by a continuous monitoring system. The coalface was staffed by three men and a deputy. It was an efficient system of coal production, but it placed much pressure on the opening out of the roadways and the coal faces to give continual production.

## Ranging drum shearers

Mainly used in thicker seams, these shearers have their discs on an arm that can be lifted and lowered. They are often used to make two passes through the coal face to complete each shear. The first pass would cut out the top layer of the seam and the powered supports would provide some forward support to the roof. On the second pass, the shearer would cut the rest of the coal seam down to floor level. This pass would be done quickly as the coal is easy to cut. As the machine made this cut, the panzer conveyor would be rammed over and the supports advanced.

**Screening run of mine coal**

It is not desirable to have to treat all the run of mine coal through the different washing processes, as coal and water and mine dirt are difficult to separate. A simple way is to feed the run of mine coal onto a wide, fixed screen which has holes, usually around one-inch diameter, drilled throughout it, and let the small coal fall onto a separate conveyor. The rest of the coal carries on to the coal preparation plant. The coal which falls through the screen is the ideal size for power station use and can be blended with other coals to achieve the required ash content.